Lecture Notes in Economics and Mathematical Systems   595

Henner Gimpel

Preferences in Negotiations

The Attachment Effect

With 33 Figures and 21 Tables

 Springer

Dr. Henner Gimpel
Karlsruhe Institute of Technology
Universität Karlsruhe (TH)
Institute of Information Systems and Management
Englerstraße 14
76131 Karlsruhe
Germany
mail@henner-gimpel.de

Library of Congress Control Number: 2007926109

ISSN 0075-8442

ISBN 978-3-540-72225-0 Springer Berlin Heidelberg New York

Springer is a part of Springer Science+Business Media

springer.com

© Springer-Verlag Berlin Heidelberg 2007

Production: LE-TEX Jelonek, Schmidt & Vöckler GbR, Leipzig
Cover-design: WMX Design GmbH, Heidelberg

SPIN 12056675   88/3180YL - 5 4 3 2 1 0   Printed on acid-free paper

Acknowledgements

This book has benefitted from contributions by many different people whom I want to thank. It would not have been possible without their help, encouragement, and ideas. First of all, I appreciate the guidance of Christof Weinhardt who has supported me in many ways; his visionary thinking and enthusiasm inspired me and my work. I'm thankful to Stefan Seifert, Matthias Burghardt and Niels Brandt for proofreading parts of the document and to Stefan and Matthias, who both have been my officemates for some years, for enduring the most stupid questions on experiments, game theory, statistics, LaTex etc. Furthermore, Carsten Holtmann, was an invaluable help in understanding sociology of research and finding motivation to pursue this work.

I'd also like to thank all other colleagues at IISM at Universität Karlsruhe (TH). Your kindness, probing questions, and expertise on engineering markets have proven extremely helpful—the lunch and coffee breaks, ZwiBis, skiing weekends, doctoral seminars, and strategy workshops have been great fun. Thanks to Björn, Carsten, Clemens, Daniel, Daniel, Dirk, Fez, Ilka, Jörg, Juho, Kiet, Matthias, and Stefan to name but a few. I'm grateful to colleagues from CIRANO and from IBM T.J. Watson Research Center for their intellectual inspiration and teaching me how to conduct experiments and write papers. Thank you Jacques, Moez, Heiko, Asit, and Bob.

Last but not least, special thanks to my family and friends who have supported and encouraged me over the last years in conducting this research.

Karlsruhe, Mai 2007   *Henner Gimpel*

Contents

List of Figures

List of Tables

1

Introduction

People make mistakes. More interestingly, people make a variety of systematic and predictable mistakes. The predictability of these mistakes means that once we identify them, we can learn to avoid them.

(Bazerman, 2006, p. 13)

Negotiations are complex, ill-structured, and uncertainty-prone processes subject to half-truths, tricks, and other means of psychological warfare (Ströbel, 2003, Ch. 2). In other words, negotiating is a demanding task with plenty of potential for making mistakes. As Bazerman (2006) points out, identifying and understanding systematic mistakes may lead to improved negotiation processes as well as facilitate the engineering of negotiation support systems. One possible systematic bias in negotiations regards attachment and the endogeneity of reference points and preferences. The following historical example illustrates the importance of endogenous reference points in negotiations, i.e. reference points that emerge in a negotiation as a results of the negotiation itself.

On September 17, 1978, Egyptian President Anwar Sadat and Israeli Prime Minister Menachem Begin signed the Camp David Accords. Prior to this agreement, Egypt and Israel had been enemies for three decades and had fought four wars. The Camp David Accords established a framework for the Egyptian-Israeli relations and led to a later peace treaty. The tense Camp David negotiations, during which US President Jimmy Carter mediated between Sadat and Begin, took thirteen days. Both parties refused to negotiate directly. On day eleven, Sadat declared he would unilaterally terminate the negotiation proceedings and leave Camp David without signing any agreement. The reason was, as reported by Carter (1982, pp. 392–393), that

> 'His own [Sadat's] advisers had pointed out the danger in his signing an agreement with the United States alone. Later, if direct discussions were ever resumed with the Israelis, they could say, "The Egyptians have already agreed to all these points. Now we will use what they have signed as the original basis for all future negotiation." It was a telling argument.'

The fear was that an intermediate step endogenously determined during the year-long negotiation process could serve as reference point for the evaluation of subsequent offers and agreements. Finally, the negotiators reached an agreement, signed the Camp David Accords and, in March 1979, a peace treaty, which was a major step in the Middle East peace process.

Carter's report explicitly illustrates the importance of reference points and their possible endogeneity to a negotiation process. Furthermore, the Camp David negotiations in particular concerned multiple issues (withdrawal from the Sinai, status of the West Bank, etc.). These two aspects—endogenous reference points and multi-issue negotiations—fall within the scope of the present work. While Sadat's advisers were aware that offers and intermediate outcomes can serve as explicit reference points, the study at hand is concerned with unconscious, systematic mistakes and biases to which a negotiator might be prone and can learn to avoid.

Besides introducing multi-issue reference points, the historical example serves to illustrate three concepts in negotiations addressed in the present work, with the following differences: Firstly, the negotiation between Israel and Egypt was a matter of international politics whereas the study at hand is concerned with commercial negotiations on the exchange of goods and services among economic entities. Secondly, the negotiation was mediated by a third party (Carter); the present work deals with negotiations in which the parties directly exchange offers. Thirdly, Sadat and Begin represented countries with diverse populations; the study at hand, however, is concerned with monolithic parties. Whether or not the results obtained in this study transfer to mediated political negotiations among non-monolithic parties, however, is beyond the scope of the present work.

In a negotiation, parties exchange offers. If these offers are mutually agreeable, an agreement may be reached. Consequently, the specific offers exchanged in a bilateral multi-issue negotiation likely influence the parties' expectations in the outcome of the negotiation.[1] Expectations in turn (unconsciously) influence the parties' reference points and whether offers and agreements are evaluated as gain or loss relative to the respective reference point. As suggested by prospect theory, gains and losses are evaluated differently, and hence, the location of a party's reference point influences her[2] preferences and trade-offs between issues. The systematic effect offers have on preferences via expectations and reference points is termed *attachment effect* in the following. The attachment effect models that a negotiator's expectation of future possession affects her attachment and obsession with possible agreements and, consequently, her concessions during a negotiation. Studying this systematic bias affecting negotiators is intended to facilitate rational negotiating and the engineering of negotiation support systems.

The attachment effect in negotiations is assessed both theoretically and experimentally in the present work. Two empirical phenomena suggest its existence and that its study may well be worthwhile: Firstly, auction fever is related to endogenous preferences in a market mechanism other than negotiations, and secondly, the rejection of Pareto improvements, i.e. changes which make one party better off after a negotiation without harming the other party, might be due to the endogeneity of preferences in negotiations.

Auction Fever

Auction fever or *bidding fever* describes the phenomenon of bidders becoming caught up in the dynamics of an auction and outbidding their initial upper limit price (e.g. Heyman, Orhun, and Ariely, 2004; Ku, Malhotra, and Murnighan, 2005). One of the possible explanations put forward for auction fever is the attachment effect: If a bidder in an auction has the highest bid for a long time, for example, she might expect to win the auction, feel that the good being auctioned 'almost certainly belongs to her possession,' and become attached to the good. If so, she perceives a loss when someone else 'takes away her good' by submitting a higher bid to the auction. As many people are loss-averse (Kahneman and Tversky, 1979), the bidder might try to regain the good by submitting higher and higher bids, thereby becoming caught up in the dynamics of the auction. Analogously, in negotiations, the attachment effect might result in a kind of *negotiation fever*: During

---

[1] The term expectation is used to denote anticipation throughout the study rather than the statistical meaning.

[2] Female pronouns will usually be used for referring to single negotiators throughout the study. In some cases, male pronouns help in differentiating two negotiators.

the negotiation process, parties could become attached to a certain element of the object of negotiation and therefore possibly perceive a loss when the counterparty proposes a trade-off which would result in the sacrifice of this element.

Rejection of Pareto Improvements

Block et al. (2006), among others, analyze data gathered with the Inspire negotiation support system. First, preferences are elicited in a pre-negotiation phase. Then the negotiation is conducted. In the event that an inefficient agreement is reached, Pareto improvements are generated by the Inspire system and presented to the negotiators. 58% of the agreements in their data set turned out to be inefficient with respect to the preferences elicited in the first phase. However, only 23% of negotiators reaching such an inefficient agreement were willing to accept the proposed Pareto improvements. At first sight, this seems puzzling and irrational, but if preferences are endogenous and change during the negotiation, the system's proposal in the post-negotiation phase may be unacceptable with respect to the ex-post preferences. This might explain the low acceptance rate. Other explanations are outlined and tested empirically by Block et al.

Vetschera (2004b) analyzes utility functions, offers made, and final agreements in a related set of Inspire negotiations. He reports that in about 25% of the cases, negotiators violated consistency in the sense that their observed behavior did not fit the ex-ante elicited utility functions. A change in preference structure predating the seemingly inconsistent behavior is one of several possible explanations. Meanwhile, nuisance in the specific utility elicitation technique employed in the first phase of negotiation support might serve to explain the observed inconsistencies as well; Vetschera presents a number of other possibilities.

Research Questions

At this juncture, several important concepts have been introduced and the focus of the study has been defined: bilateral commercial multi-issue negotiations and changes in the negotiators' preferences that may be triggered by attachment and reference points. More precisely, the following four questions guide the subsequent analysis:

1. Are preferences endogenous to negotiations, i.e. are they influenced by the specific course of a negotiation?
2. Can models that allow for endogeneity of preferences predict behavior significantly better than models relying on exogenous preferences?

3. Is there a systematic bias of preferences depending on the offers exchanged in a negotiation?
4. If it is the case that preferences are reference-dependent, how is the reference point determined?

Delving into these questions is worthwhile because the answers have implications for preparing and conducting negotiations as well as for the engineering of negotiation support systems. The first question implicitly challenges the applicability of traditional economic rational choice models for understanding and conducting negotiations. If it can be answered affirmatively, the next step is to compare the models listed in the second question vis-a-vis their ability to predict negotiator preferences. The third question builds upon the first by raising the possibility that the offers exchanged in a negotiation might play a pivotal role; it also provides the implicit foundation for building the model called for in the second question. Finally, the fourth question adds another hypothesis as to how offers might influence preferences, namely via reference points. Answers to the four questions will be presented in the concluding Chapter 6 based on a theoretical and experimental analysis.

1.1 Related Work and Fields of Research

Work related to the study of preferences in negotiations comes from various fields, most prominently from negotiation analysis and behavioral decision research, prospect theory, game theory, cognitive psychology, and information systems research. The purpose of this section is twofold: Firstly, it relates the present work to previous literature and indicates how the different aforementioned fields influence the following study. (At this point, numerous references will be cited without going into detail; a more in-depth discussion will follow in the next chapters.) Secondly, the section presents a detailed discussion of the two most closely related studies: experiments by Kristensen and Gärling (1997a) and Curhan, Neale, and Ross (2004).

Negotiation Analysis

Negotiation analysis integrates (behavioral) decision sciences and game theory to bridge the discrepancy between descriptive and prescriptive approaches to negotiations. This field of research was initiated by Raiffa (1982) who proposed an *asymmetric prescriptive/descriptive approach* in his seminal book on the art and science of negotiation. Descriptions of behavior in negotiations compiled largely from research in psychology, behavioral economics, and experimental economics form the basis for advising negotiators on how to negotiate rationally.

A core element of the descriptive basis of negotiation analysis is a set of common biases in negotiations. These biases predict how decision-makers' cognition and behavior systematically deviates from prescriptive models. One example is the famous *fixed pie illusion*: It was found that negotiators often disregard the integrative potential of multi-issue negotiations and assume that their preferences strictly oppose their counterparty's preferences. Thus, they focus on competitive issues, and as a result, agreements are frequently either inefficient or unable to be reached at all (Bazerman, Magliozzi, and Neale, 1985; Thompson and Hastie, 1990; Thompson and DeHarpport, 1994; Fukuno and Ohbuchi, 1997). Other common biases in negotiations are anchoring and adjustment (e.g. Northcraft and Neale, 1987), framing (e.g. Bazerman, Magliozzi, and Neale, 1985), the availability bias (e.g. Pinkley, Griffith, and Northcraft, 1995), overconfidence (e.g. Kramer, Newton, and Pommerenke, 1993), the illusion of conflict (e.g. Thompson, 1990), reactive devaluations (e.g. Ross and Stillinger, 1991), escalation of conflict (e.g. Bazerman and Neale, 1983), ignorance of the other's behavior (e.g. Bazerman and Carroll, 1987), and egocentrism (e.g. Camerer and Loewenstein, 1993); see Section 3.1.3 for a review of these biases in negotiations as well as a more extensive bibliography. Further collections of common biases used in negotiation analysis are provided by Neale and Bazerman (1991, Ch. 3 & 4), Bazerman and Neale (1992, Part I), Bazerman et al. (2000), and Bazerman (2006, Ch. 10).

The present work identifies the attachment effect as an additional, novel bias in negotiations. The different biases and studies do not contradict each other, but together constitute a large part of the descriptive basis on which negotiation analysis builds.

So far, the references to common biases have served to position the present work in the context of negotiation analysis; they will be discussed in greater depth in Chapter 3. In the following paragraphs, the two studies most closely related to the attachment effect are examined in more detail.

Endogenous Preferences in Negotiations

In terms of research on negotiations, reference-dependent evaluation of offers is not new. Most studies do, however, assess static exogenous reference points like market prices or reservation prices (Kahneman, 1992; White et al., 1994). A reservation price is the price beyond which a negotiator would prefer not reaching an agreement at all (Raiffa,

1982, Ch. 4). The present work analyzes how reference points are endogenously determined in the process of negotiating; in this regard, it is most closely related to the work by Kristensen and Gärling (1997a) and Curhan, Neale, and Ross (2004).

Kristensen and Gärling (1997a) study the selection process of one of several possible reference points in single-issue negotiations on the price of condominiums that subjects hypothetically consider for purchase. As part of their experiment, they induce a reservation price and assume that it might function as an exogenous reference point. Hence, prices might be evaluated as gains or losses from this reference price. Meanwhile, they regard the seller's initial offer as a second possible reference point. This second reference is endogenous to the negotiation process. In a series of experiments, Kristensen and Gärling vary the values and relative location of these two possible reference points and analyze which of the two affects offers by having their subjects play the role of buyers. They find that most commonly, the sellers' initial offers are adopted as reference points by buyers. In some settings, however, the exogenous market price is influential as well. The authors conclude that although negotiators take various pieces of information into account, there is no single dominant reference point (as e.g. suggested by White et al., 1994).

Two aspects about the work by Kristensen and Gärling are especially noteworthy here: Firstly, the authors provide evidence that offers might be adopted as reference points. In this respect, their results are in congruence with the present work. In fact, their findings can be explained by the attachment effect model, which will be discussed in Section 3.3. Secondly, the authors assume that the adoption of a reference point is all-or-none (e.g. Kahneman, 1992). Under this premise, Kristensen and Gärling describe difficulties in explaining why different pieces of information affect reference points. They speculate that either subjects switch from one reference point (e.g. the reservation price) to another (e.g. the initial offer) over time, or that the measured effect reflects an 'average' of the subjects' various reference point. This problem can be alleviated by explaining their results with the attachment effect. The adoption of reference points is not in fact all-or-none (Strahilevitz and Loewenstein, 1998). The attachment effect allows that different pieces of information enter into a negotiator's expectations with respect to the outcome of the negotiation and thus influence reference points. How strongly information affects reference points depends on the reliability of the information and the negotiator's subjective judgment on how relevant it is for the final agreement.

In contrast to the work by Kristensen and Gärling (1997a) described above, the present study will concern multiple issues, involve an alternating offer exchange rather than just a single offer, will not be solely hypothetical, and will not assume that the adoption of a reference point is all-or-none.

Curhan, Neale, and Ross (2004) experimentally study changes in negotiators' preferences during a multi-issue negotiation with a focus on dissonance and reaction theory. The two main hypotheses are as follows: Firstly, cognitive dissonance theory (Festinger, 1957; Festinger and Aronsons, 1960) suggests that decision-makers tend to reduce discrepancies that might exist between different cognitive patterns. In the context of multi-issue negotiations, this means that a negotiator might feel more positively about an offer once she proposed it to her counterparty as potential agreement. Secondly, reaction theory (Brehm, 1966) suggests a reactive devaluation of any offers received from the counterparty. In the experiment by Curhan et al., subjects bilaterally negotiated on three issues of a student loan contract; in each round, both negotiators simultaneously write down offers, then rate each potential agreement with respect to its desirability, and finally have two and a half minutes to argue, explain, etc. Agreement is reached when both parties write down the same offer in any given round.

Curhan et al. indeed find evidence that their subjects' preferences were influenced by the offers exchanged. As dissonance theory suggests, subjects tended to express higher preference for contracts they themselves had offered. This tendency was even stronger when a contract became the final agreement. The evidence for reactive devaluation is inconclusive.

In the present context, two aspects of the study by Curhan et al. are worthy of mention: Firstly, the experiment suggests—as does the present study—that preferences might change endogenously over the course of a multi-issue negotiation dependent on the offers exchanged. Secondly, Curhan et al. attribute the increased preference for offers a negotiator has proposed herself and for agreements to dissonance; it can, however, also be explained by the attachment effect. This will be discussed in Section 3.3 after the attachment effect is described in more detail.

Prospect Theory

A core element of the attachment effect is the dependence of preferences on reference points. With respect to the implications of reference points, the present work is most closely related to a study by Tversky

and Kahneman (1991), who extend the concepts *reference dependence*
and *loss aversion* from risky choices (Kahneman and Tversky, 1979)
to riskless multi-issue choices. The field study by Hardie, Johnson, and
Fader (1993) is one of the few works to empirically test the existence
of issue-wise reference points and their implications on trade-offs between issues. The significance of (exogenously given) reference points
for negotiations is discussed by Kahneman (1992).

With respect to the origin of reference points, the present work
draws on the traditional view that the status quo of property rights
determines the reference point as well as on the alternative interpretation that expectations are essential in determining the location of a
reference point. To this end, the attachment effect is related to the endowment effect (e.g. Knetsch, 1989; Kahneman, Knetsch, and Thaler,
1990; Camerer, 2001), the history of ownership effect (Strahilevitz and
Loewenstein, 1998) and the role of expectations (Köszegi and Rabin,
2006). Furthermore, the attachment effect in negotiations is comparable
to the explanation of auction fever via attachment or quasi-endowment
(e.g. Heyman, Orhun, and Ariely, 2004; Ku, Malhotra, and Murnighan,
2005; Abele, Ehrhart, and Ott, 2006).

Game Theory

The present work neither introduces nor uses a game theoretic equilibrium model. While game theoretic modeling would generally be a
conventional methodology for assessing the strategic interaction of negotiators, it is not pursued in the present study as there are infinitely
many Nash equilibria in alternating offer multi-issue negotiations under incomplete information, and (to date) no meaningful refinement of
this equilibrium concept exists capable of singling out a small set of
equilibria or a set of reasonably homogenous equilibria (cf. Sec. 3.1.1).[3]

Nevertheless, the present work is related to game theoretic studies of
bilateral single-issue bargaining, most closely to the models by Shalev
(2002), Li (2004), Hyndman (2005), and Compte and Jehiel (2006). All
these authors incorporate reference-dependent preferences into more
or less standard models of bargaining, assume that the reference point
is endogenously determined in a negotiation, and identify equilibrium
strategies and characteristics of equilibria. The formalization of the
attachment effect that will be presented in Chapter 3 was inspired

---

[3] Note that a negotiation analytic approach would be appropriate even if a unanimous equilibrium could be derived in theory. Game theory builds on the rationality of all players whereas negotiation analysis aims at more pragmatic advice
on how to negotiate in the absence of 'hyper-rational' agents.

by (an earlier version of) the model by Compte and Jehiel (2006). See Section 3.3.3 for a more detailed discussion of these single-issue equilibrium models and how they relate to the attachment effect.

Information Systems

Electronic communication media in negotiations, negotiation support systems, and automated negotiations are studied in computer science and information systems research. Research in this field includes process models for negotiations that relate to the present study. Most notably, these are the media reference model by Schmid (1999) and Lechner and Schmid (1999), parts of the Montreal Taxonomy by Ströbel and Weinhardt (2003), and a distinction in private and shared information in negotiations proposed by Jertila and Schoop (2005). These process perspectives are employed to identify sub-processes in negotiations and correlate them to mental processes in which negotiators might be prone to biases (cf. Sec. 3.1). Furthermore, there is a correlation between the information systems research literature (especially on negotiation and market engineering) and the study of negotiator perception and behavior (e.g. Köszegi, Vetschera, and Kersten, 2004, and Lai et al., 2006).

Related Publications

Design and results of the internet experiment (cf. Ch. 4) have been accepted for publication by the *Group Decision and Negotiation* journal (Gimpel, 2007). The results described there are integrated into the present work to give a comprehensive account of the attachment effect and the experimental evidence. Furthermore, drafts of and ideas from the present work have been presented at various conferences: Dagstuhl Seminar 'Computing and Markets' 2005, Schloss Dagstuhl, Germany (Gimpel, 2005); Annual Meeting of the Gesellschaft für Experimentelle Wirtschaftsforschung 2005 (GEW 2005), Cologne, Germany; Annual Meeting of the Economic Science Association 2005 (ESA 2005), Montreal, Canada; Group Decision and Negotiation 2005 (GDN 2005), Vienna, Austria; Dagstuhl Seminar 'Negotiation and Market Engineering' 2006, Schloss Dagstuhl, Germany (Gimpel, 2006).

1.2 General Background

The general background of the present work is twofold: On the one hand, it is situated in research on *negotiation analysis*; on the other hand, it belongs to the field of *negotiation and market engineering*.

Negotiation analysis, with its asymmetric prescriptive/descriptive approach to analyzing negotiations and advising parties how to negotiate rationally, was already introduced as related work in the previous section. The focus of negotiation analysis is on the individual parties in a negotiation (individual decision-makers, non-monolithic parties, mediators, arbitrators, etc.) and their behavior. On the contrary, the focus of *negotiation and market engineering* is on the design of institutions and systems that structure the interaction of individuals and organizations.

In recent years, economics has exhibited a tendency to partially evolve from a positive science to an applied engineering discipline. In positive economics, researchers develop and verify (abstract) theories that explain and predict human and organizational behavior. In contrast, economic engineering is often called a design science, the art of economics, or applied policy analysis. Economic engineers correlate insights from positive economics to real world problems and situations. They create new and innovative artifacts to extend the limits of human and organizational capabilities (Colander, 1994; Hevner et al., 2004; Gimpel and Mäkiö, 2006).

The engineering of the FCC spectrum auctions in the US (e.g. McAfee and McMillan, 1996), the job market for graduates in medicine (Roth and Pearson, 1999), and the electric power market in California (Wilson, 2002) all teach an important lesson: It is difficult to comprehend an economic or social system unless one can intervene and experiment with it. It is even more difficult to predict a system's future behavior, unless it has been shaped and engineered so as to work 'appropriately' (Guala, 2005, Ch. 8). Another area of recent development that clearly underscores the necessity of engineering markets and negotiations is the increasing presence and relevance of electronic markets. While in traditional physical markets the rules might evolve over time, electronic markets make the conscious and structured design of the rules of interaction indispensable, as they have to be implemented in computer systems and do not allow spontaneous changes. Smith (2003) points out the necessity of a structured approach to engineering markets in his Nobel Prize lecture by stating that 'all worthwhile social institutions were and should be created by conscious deductive processes of human reason' (pp. 504–505). A predominant domain where economic engineering has been applied in the last decade is market design (Roth, 2002; Varian, 2002); Weinhardt, Holtmann, and Neumann (2003) coined the term *market engineering* to denote the conscious, structured, systematic, and theoretically founded procedure of analyzing, designing and introducing electronic market institutions. See also

Neumann (2004) and Weinhardt and Gimpel (2006) for a more extensive discussion.

The argument for the conscious design of market institutions such as financial exchanges, spectrum auctions, and electricity markets likewise applies to engineering (electronic) negotiations. Negotiations are ubiquitous; electronically supported negotiations have become essential for business life over the past few decades. Recent years have witnessed significant changes in electronic markets and trading organizations enabled by new technologies. These new technologies have created substantial opportunities for negotiation support and automated trading. The design of systems that are easy to use and can satisfy negotiators' requirements reflects the *negotiation engineering* approach (Kersten, 2003; Ströbel, 2003). This engineering approach utilizes results from positive economics and other disciplines to find solutions to practical problems.

The major challenge in negotiation and market engineering is to assess the behavior of the participating agents: How will they respond to a given economic institution (an auction mechanism, negotiation protocol, etc.), IT infrastructure, or market operator business structure? How will their behavior be affected by the socio-economic and legal environment in which these entities are embedded? Different tools from various disciplines are used to assess agent behavior (Weinhardt, Holtmann, and Neumann, 2003; Kersten, 2003; Ströbel, 2003; Weinhardt and Gimpel, 2006); among these are (game) theoretic modeling, computer simulations, field studies, and experiments. The present work is situated in this research on negotiation and market engineering: It studies the preferences and behavior of negotiators, thus contributing to the positive basis upon which negotiation engineers can devise protocols and systems to assist negotiators; these can then be deployed to engineer other market institutions.

1.3 Overview and Structure

The structure of the work at hand is schematized in Figure 1.1. After the present introduction to the context of this work, Chapter 2 presents and compares several theories on preferences as a first approach to a theoretical understanding of the cognition and behavior of negotiators. This comprises traditional microeconomic theory, behavioral economics, cognitive psychology, and the neuro-sciences.These approaches to human decision-making differ with respect to internal coherence, congruence with reality, abstraction, and predominant research methodologies. There is no overall best theory on preferences for studying multi-issue negotiations. Behavioral economics and reference-dependent preferences are, however, most important for the subsequent chapters.

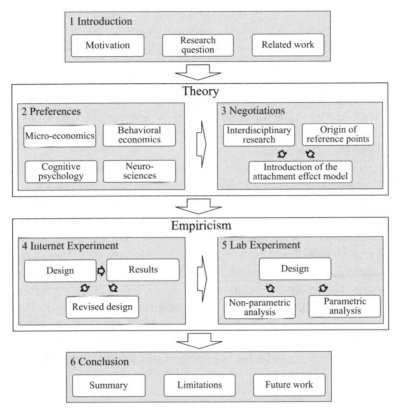

Fig. 1.1. Structural overview

Chapter 3 then investigates the specifics of negotiations and decision-making in negotiations. It starts by sketching the interdisciplinarity of research on negotiations. The focus of the presentation centers on game theoretic models that correspond to the microeconomic theory on preferences presented in Chapter 2 and on negotiation analysis related to the behavioral perspective on preferences. Based on the general assumption that preferences are defined relative to reference points, the origin of such reference points is discussed. The traditional supposition is that the status quo serves as a reference point, whereas more recently, the role of expectations has gained prominence. Based on expectations, the attachment effect in negotiations is exemplified, modeled, and discussed in Section 3.3.

Chapter 4 empirically tests for the existence of the attachment effect that was introduced theoretically in the previous chapters. The design and the results of an internet experiment on multi-issue negotiations between human subjects and a software agent is reported. In the experiment, negotiators exchange offers on the terms of a (hypothetical) tenancy contract and are subsequently asked to judge the importance of single issues in the contract. The data suggests that negotiators' preferences are systematically biased by the attachment effect. At the end of the chapter, the design is revised by lessons learned from the experiment to rule out possible concerns regarding the validity of results in a follow-up experiment.

Chapter 5 presents a second experiment to reinforce both the internal and external validity of the results from the internet experiment. The experiment is conducted in the lab with salient rewards and—as the first experiment—this second experiment favors the attachment effect model over a traditional rational choice model. This is shown by several non-parametric statistics and an estimation of the parameters in the attachment effect model allows the quantification of the effect of single offers on reference points.

Chapter 6 concludes the work by summarizing the results and contributions to research on negotiations. It critically discusses the limitations of the present work and indicates directions for possible future work on endogenous preferences in negotiations and markets in general.

2

Theories on Preferences

I have said on another occasion, and it seems to me important enough to repeat it here, that he who is only an economist cannot be a good economist. Much more than in the natural sciences, it is true in the social sciences that there is hardly a concrete problem which can be adequately answered on the basis of a single special discipline.

(Hayek, 1967, p. 267)

Since negotiators are decision-makers, understanding a negotiation requires a deep understanding of the negotiators' decisions. As Hayek suggests, the theoretical foundations in this chapter address decision-making and preferences from the viewpoint of different disciplines. The origin of preferences and their stability over time varies widely across fields: Economists, for example, usually assume preferences to be an underlying property of any individual and to be stable over time. If an agent's choice changes over time, then either the production technology available or the information at hand have changed—preferences do not. This widely used perspective is most notably vindicated by Stigler and Becker (1977) in a seminal paper arguing against the assumption of changing preferences and it is outlined in several microeconomic textbooks, e.g. Kreps (1990), Varian (1992), Mas-Colell, Whinston, and Green (1995).

Another perspective on preferences takes a more psychological view: Preferences are constructed by the time an agent faces a choice situation. In this perspective, preference construction is a mental process highly dependent on the context of the decision environment. Therefore, preferences are not (necessarily) stable over time—preferences change along with the context. The context includes, for example, the framing of a decision as winning or loosing, the arrangement along with other choices, and the social situation. This second perspective on preferences is frequently used in psychology and behavioral economics, e.g. by Payne, Bettman, and Johnson (1993), Bettman, Luce, and Payne (1998), Hoeffler and Ariely (1999), and Ariely, Loewenstein, and Prelec (2003).

The topic of the present work is to study decision-making in negotiations. Preferences are the traditional economic modeling device to assess what decision-makers—like negotiators—want to achieve and how they compare different possible outcomes. Thus, they are essential to understand behavior in negotiations. However, as briefly sketched in the last paragraphs, there are different theories on preferences and their properties. To shed more light on these different theories, this chapter reviews microeconomic and psychological approaches to human decision-making. The remainder of the preamble to this chapter is devoted to introducing some terminology and to present a rough classification of the different approaches along three dimensions: (1) their degree of abstraction, (2) the predominant research methodology, and (3) the underlying theory of truth. This classification intends to show the usefulness of the coexistence of different theories on preferences and the fact that a study can gain by drawing on different theories. Reviewing specific theories on preferences then starts with the traditional microeconomic view presented in Section 2.1.

Approaches to Human Decision Making

Economic theory builds abstract, oftentimes mathematical models of the real world. Like any model, economic models reduce the complexity of the real world by simplifying assumptions like the rationality of agents. The aim of economic theory is to clarify the connections between different types of concepts, arguments, and patterns of reasoning. Economic theorists (usually) do not claim that their assumptions are descriptively valid. Their purpose is not to model individual decision-makers as close to reality as possible; it rather is to make reasonable simplifying assumptions so that models highlight the interrelation of important economic concepts and institutions and, furthermore, that their suppositions should on aggregate correspond to reality. A good economic model is realistic in the sense that it orders perception of real life phenomena (Rubinstein, 2001).

Experimentation has become a widely accepted methodology in economics over the last decades. Friedman and Sunder (1994, Ch. 1), for example, point out that while the theory organizes our knowledge and enables us to predict behavior in new situations, experimentation oftentimes sheds light on regularities that are not (yet) explained by existing theory. Such empirical regularities stimulate refinement of theory.

Besides theoretical and experimental economics, human decision-making is approached by psychology. Psychology—the study of the mind, brain, and behavior—is an inherently empirical, descriptive discipline: the purpose is not to build abstract models, but to understand how a single individual (embedded in a social context) perceives the world and makes decisions. The field most closely related to decision-making in negotiations is cognitive psychology; it studies the mental processes of cognition, information processing, and behavior.

Finally, and even less abstract, neuro-sciences study the neural basis of mental processes like perception and information processing. Knowledge of how the human brain interacts with its environment allows researchers to gain a new perspective on the variation within and between individuals' decision-making. This in turn allows—according to McCabe (2003)—to better predict economic behavior and engineer institutions to structure social and economic interactions.

Theories of Truth

Different disciplines differ with respect to the underlying theories of truth and it is helpful to briefly distinguish them in order to compare approaches to human decision-making. Obviously, any scientific discipline aims at truth of statements, theories, theorems, propositions, etc. But how exactly is *truth* defined? What does it mean that a theorem is true? The answer to these questions depends on the theory of truth a discipline adopts. The two theories of truth considered in the following are *coherence theory* and *correspondence theory*. According to coherence theory, the truth of a 'proposition can only be determined by establishing whether it coheres with other propositions we already know to be true.' (Bush, 1993, p. 69); a proposition is true if it can be embedded in a consistent, comprehensive system of other propositions that are related by the terminology used and by logic associations.[1] If coherence theory is used as a measure, a scientific theory does not at all have to be related to the real world as long as it is internally

---

[1] Logical consistency is one interpretation of coherence theory. Some philosophers add further requirements like coherence with beliefs by the researcher herself, by the majority of people in a society, etc.

consistent. Mathematical modeling and reasoning, as used in economic theory, are convenient tools to achieve internal coherence.[2]

According to correspondence theory of truth, 'the truth of a proposition (which is subjective) is established if it "corresponds" to the facts (which are objective).' (Bush, 1993, p. 67) This understanding requires that there is a 'reality', and hence, a non-constructivist world view like naive realism or critical realism is assumed. In psychology or neuro-sciences, a theory is not judged highly for its abstract elegance, but for its resemblance of the true nature of human decision-making.

While some philosophers of science treat coherence and correspondence as two rival accounts of the concept of truth, McCloskey (1994, p. 276) points out the necessity of both: 'The trouble with correspondence is that without coherence it applies only to simple cases [...] likewise coherence without correspondence is not much. It's just chitchat, or mathematics. So we need in science and in life both coherence and correspondence.' Correspondence theory focuses on empiricism and studies decision-making in relation to the way the world works; on the contrary, coherence theorists are interested in decision-making in relation to the way it ought to work (Connolly, Arkes, and Hammond, 1999, Ch. 3).

Interdisciplinarity

The different approaches to human decision-making—economics, psychology, neuro-sciences—are not mutually exclusive and not in contradiction to one another. The borderline between economics and psychology, for example, is not a sharp distinction but a rather wide field. Over the last decades, the term *behavioral economics* was coined to denote research in this area. Furthermore, many psychologists collaborate with neuro-scientists to better understand the biological processes in the brain (this is termed *neuro-psychology*). The mix of economics and neuro-sciences which became increasingly popular in recent years is oftentimes referred to as *neuro-economics*.

Figure 2.1 relates the different approaches to human decision-making. Besides the disciplines economics, psychology, and neuro-sciences, behavioral economics is sketched as well to highlight its prominent position in this work. The overlapping boxes indicate the fluid transition from one field to the other. The approaches to decision-making differ with respect to (1) the degree of abstraction, (2) the

---

[2] The relation of economic theory to the real world is not denied here—it is, however, not necessary for coherence theory.

main research methodology, and (3) the relative importance of underlying theories of truth. This is indicated by the labels at the side: coherence theory of truth, for example, is increasingly important as one goes to the upper left, i.e. to economics, and correspondence theory is increasingly important if one goes to the lower right, i.e. to neurosciences.

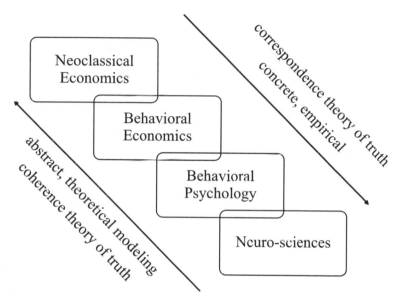

Fig. 2.1. Approaches to human decision-making

Neoclassical economics is an abstract theory that heavily relies on theoretical, mathematical modeling. The core model is optimization subject to constraints that arise from institutional rules, resource limitations, and/ or the behavior of others (Smith, 2003). The traditional, predominant methodology is theoretical modeling. Besides that, an increasing amount of empirical work in the field and the lab is done around these theoretical models.

The farther one goes to the lower right end of Figure 2.1, the more concrete and the more empirical the approaches become. This can be seen from an example on decisions under risk: The most prominent concept in behavioral economics is prospect theory whereas expected utility theory is the traditional neoclassical pendant. Prospect theory is less abstract than expected utility theory as it was inspired by empirically observed violations of the latter and tries to fit the data more closely. Nevertheless, prospect theory is to a wide extent a utility optimization theory; it is far more abstract than neuro-imaging studies locating different brain regions that are activated for processing information on probabilities and outcomes.

Figure 2.1 does not provide exact boundaries between the different approaches to decision-making based on either the theory of truth, the level of abstraction, or the predominant methodology. In each of the approaches, there is some coherence and some correspondence, some abstraction and some concreteness, some theoretical modeling and some empiricism. The relative weighting of these concepts does, however, differ between disciplines.

Each of the above approaches to decision-making and preferences has its strengths and weaknesses when it comes to understanding and predicting the behavior of negotiators. There is no unanimously 'best' or 'worst' theory, as this would inevitably require trading off the relative importance of a theory's characteristics like generality, manageability, tractability, congruence with reality, and predictive accuracy. A more general theory is better, a theory with higher manageability is better, a theory in congruence with reality is better, etc. (Stigler, 1950) but if there is no theory that is best with respect to each of these dimensions—and there is no such theory for individual decision-making in negotiations—then quality of a theory depends on the subjective weighting of these characteristics and integrating different approaches becomes beneficial.

Outline of the Chapter

The remainder of the chapter is devoted to introducing the different approaches in more detail as concepts and results from each of them are used for the subsequent study of negotiators' preferences and behavior. To this end, Section 2.1 reviews a standard microeconomic model of preferences, utility, and rational choice. Afterwards, the presentation turns to behavioral economics, especially to prospect theory as an approach which takes the decision-makers' cognition into account (Sec. 2.2), and then to the psychological perspective which assumes that preferences are constructed on the fly (Sec. 2.3). Section 2.4 briefly reviews some interesting neuro-scientific results on decision-making and, finally, Section 2.5 sums up the discussion, relates the different approaches to one another and points out some strengths and weaknesses.

Terminology

The terminology employed when talking about preferences is diverse. The present analysis uses the term *agent* to denote an individual who makes a decision; other authors use decision-maker, consumer, person, or individual instead. In subsequent chapters the term *software agent* will be used to denote a computer program that makes a decision based on a pre-specified decision rule. (Agent is used as short version of software agent if the agent's nature is clear.)

If there are several objectives an agent is interested in, these are called *issues*. Other terms used in the same context are attributes, dimensions, features, objectives, criteria, and aspects.

Alternatives refer to the different options an agent can choose from—outcome, option, prospect, and action are different terms for this concept. Some authors would differentiate all options in two categories: a single option is the considered option and all other ones are the alternatives. However, in the present study *all* options potentially chosen are referred to as alternatives.

2.1 Microeconomic Modeling of Preferences

Preferences are a central modeling device in microeconomics. Together with utility as numerical representation of preferences they allow to capture what an individual agent desires in a relatively easy and analytically tractable way. Normative theory of decision-making applies preferences as *reason for behavior*, i.e. preferences capture an agent's characteristics other than beliefs and capacities. In a descriptive theory, on the other hand, it is not necessary that agents really possess preferences and deliberately adhere to them. Rather, it is sufficient that they act *as if* guided by preferences and utility maximization.

The notion of stable preferences is picked up by exemplifying a standard rational choice model in Section 2.1.1. Subsequently, the descriptive validity of the underlying assumptions is challenged in Section 2.1.2.

2.1.1 Rational Choice Theory

Neoclassical theory takes preferences as exogenously given and fix and they are assumed to be invariant to the elicitation procedure as well as agnostic to (market) processes. Formalization in line with this perspective on the nature of preferences is outlined in the following.

In almost every microeconomic approach, the standard way of modeling an agent, i.e. the consumer or the decision-maker, is via preference relations. The following exposition of this standard tool is mainly based on Kreps (1990, Ch. 2). Proofs are thereby omitted; see e.g.

Kreps (1988) for the proofs of propositions or (Fishburn, 1970) for a more rigorous mathematical introduction.

Let X be a finite set of (mutually exclusive) alternatives an agent can choose from.[3] Such a set can, for example, contain four alternatives: 'buy nothing', 'buy good A', 'buy good B', or 'buy both'. If for two alternatives $x, y \in X$ the agent says that x is better than y, this is denoted by $x \succ y$, i.e. x is strictly preferred to y. A binary relation $\succ$ on X is termed preference relation if it is *asymmetric* and *negatively transitive*.[4]

Asymmetry

Asymmetry of strict preferences states that it is impossible for an agent to (strictly) prefer alternative x over y and—at the same time—prefer y over x. If either alternative is strictly better, then the other one can not be better as well.

Definition 2.1 (Asymmetry of preferences). *Strict preferences are asymmetric if $\nexists\, x, y \in X : x \succ y$ and $y \succ x$.*

As an example let the alternatives the agent faces be as follows: If the agent chooses x, she receives € 200; if she chooses y, she gets € 100. Assume furthermore that the agent judges more money to be better than less money. Thus, she prefers x to y, i.e. $x \succ y$. Asymmetry then requires that the agent does not think at the same time that € 100 is better for her than € 200.

Negative Transitivity

Besides being asymmetric, preference relations are negatively transitive. Negative transitivity is somewhat trickier than asymmetry. Reconsider the example of an agent choosing among $x = $ € 200 and $y = $ € 100. What happens if a new alternative z is inserted into the choice set X? Negative transitivity of preferences requires that the agent can compare it via strict preference to at least one of the previous alternatives if one of them is strictly preferred to the other. If the new alternative is receiving € 300, for example, preferring more over less money implies $z \succ x$ (and $z \succ y$ at the same time). No matter which amount of money alternative z offers, it can be strictly compared to at least one of € 100 or € 200.

---

[3] Note the different terminology employed by several authors as outlined at the beginning of the chapter.

[4] Several other presentations of modeling preferences start with weak preference and indifference. These relations will be introduced later on.

Definition 2.2 (Negative transitivity of preferences). *Preferences are negatively transitive if* $\forall\, x, y, z \in X : x \succ y \Rightarrow x \succ z \lor z \succ y,$ *or both.*[5]

With multi-issue choices, negative transitivity is not as straightforward as the example above. A multi-issue choice is a choice in which the alternative has implications on several issues which are important to the agent. Issues can be different attributes of a contract, different goods to be purchased, or any other characteristic of an alternative outcome. In a two-issue scenario, for example, the agent could be interested in money she receives in two different currencies—dollars and euros. Continue to assume that the agent prefers more to less money in either currency. Figure 2.2 schematically sketches possible alternatives to choose from; the two dimensions labeled issue 1 and issue 2 represent the different currencies in this example.

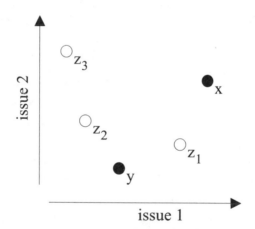

Fig. 2.2. Example for alternatives in a two-issue choice situation

Alternative x in Figure 2.2 offers more in either currency than alternative y; the strict preference is $x \succ y$. Introducing z_1 is straightforward: the two strict comparisons $x \succ z_1$ and $z_1 \succ y$ both hold—negative transitivity is satisfied. Alternative z_2 can not easily be compared with y, but with x $(x \succ z_1)$. According to Definition 2.2, one comparison is sufficient and, thus, negative transitivity is satisfied.

---

[5] The name negative transitivity becomes clear with a minor transformation: $\forall\, x, y, z \in X : x \succ y \Rightarrow x \succ z \lor z \succ y \Leftrightarrow x \not\succ z \land z \not\succ y \Rightarrow x \not\succ y$ which is the usual definition of transitivity with a negation of the preference relation. However, the definition given above is more convenient in the present context.

Alternative z_3 is the tricky one as there is no dominance relation to either x or y. On issue 2, the agent prefers z_3 to both original alternatives x and y. On issue 1, on the other hand, z_3 is the worst alternative. The agent has a conflict of interest in the two issues. This is the case where requiring negative transitivity has real significance as it implies that the agent can trade off a gain in one issue against a loss in another issue.

It is exactly this trade-off across issues and its stability during a negotiation that is discussed in the following chapters. Neoclassical models—as outlined here—assume the trade-off to be exogenously given and fix. On the contrary, behavioral approaches argue that it might change over time, for example, as a consequence of a negotiation.

In the scenario with two different currencies outlined above, a trade-off seems natural as the agent could, for example, use the exchange rate in the monetary market to convert one currency in the other. For other issues, the existence of well defined trade-offs might not be so obvious: Imagine for example a consumer trading off a car's safety against its comfort or assume a legislation process in which threats to human lives have to be traded-off against costs of prevention.

Additional Properties of Strict Preference

Besides asymmetry and negative transitivity, strict preferences show other properties that appear natural. They are listed below for completeness of the presentation. All of these properties follow from the fact that the strict preference relation is asymmetric and negatively transitive:[6]

- *Irreflexivity* assures that no alternative is strictly preferred to itself ($\nexists\, x \in X : x \succ x$).
- *Transitivity* denotes that an agent preferring x over y and y over z will prefer x over z as well ($\forall\, x, y, z \in X : x \succ y \wedge y \succ z \Rightarrow x \succ z$).
- *Acyclicity* directly follows from irreflexivity and transitivity. It represents the fact that if an agent prefers an alternative x_1 over x_n—potentially established via a chain of strict preferences—then the two alternatives are not the same (for any finite integer n: $x_1 \succ x_2, x_2 \succ x_3, \cdots, x_{n-1} \succ x_n \Rightarrow x_n \neq x_1$).

Weak Preference and Indifference

Weak preference ($\succeq$) is defined as absence of strict preference in the opposite direction ($\forall\, x, y \in X : x \succeq y \Leftrightarrow \neg(y \succ x)$), and indifference

---

[6] The proof is provided by Kreps (1988, p. 9) for his proposition 2.3.

($\sim$) as the absence of strict preference in either direction ($\forall\, x, y \in X$: $x \sim y \Leftrightarrow \neg(x \succ y \vee y \succ x)$).

Following from above properties of strict preference, i.e. asymmetry and negative transitivity, weak preference $\succeq$ is[7]

- *complete* ($\forall\, x, y \in X$ either $x \succeq y$ or $y \succeq x$, or both) and
- *transitive* ($\forall\, x, y, z \in X : x \succeq y \wedge y \succeq z \Rightarrow x \succeq z$)

  and indifference $\sim$ is

- *reflexive* ($\forall\, x \in X : x \sim x$),
- *symmetric* ($\forall\, x, y \in X : x \sim y \Rightarrow y \sim x$), and
- *transitive* ($\forall\, x, y, z, \in X : x \sim y \wedge y \sim z \Rightarrow x \sim z$).

Several textbooks start their introduction on consumer preferences with the weak preference relation $\succeq$ and its properties (e.g. Varian, 1992; Mas-Colell, Whinston, and Green, 1995). From there on, they define strict preference. This approach yields results equivalent to the ones presented here by starting with asymmetric and negatively transitive strict preference.

Utility as Numerical Representation

The binary relations $\succ$, $\succeq$ and $\sim$ introduced so far allow pairwise comparisons of alternatives an agent can choose from. This can be represented numerically for convenient handling.

Definition 2.3 (Utility function). *Given a preference relation $\succ$ on a finite set of alternative choices X, a utility function for those preferences is any function $u : X \Rightarrow \mathbb{R}$ such that $\forall\, x, y \in X : x \succ y \Leftrightarrow u(x) > u(y)$.*

A utility function $u(\cdot)$ assigns a (numerical) label to any set of alternatives for which the agent is indifferent. Furthermore, a higher number indicates that the respective alternative is strictly preferred to an alternative with a lower number. Such a numerical representation exists whenever X is finite and $\succ$ is asymmetric as well as negatively transitive.[8] The finiteness of X is not a necessary condition under some other assumptions. However, for the present study the treatment of a finite set of alternatives is sufficient.

---

[7] The proof is provided by Kreps (1988, p. 10) for his proposition 2.4.

[8] The proof is provided by Kreps (1988, pp. 19–22) for his proposition 3.2.

Ordinal Utility

Defined as above, utility is an ordinal measure and there are infinitely many utility functions to represent a given preference ordering. An ordinal utility function is unique only up to strictly increasing transformations. With an ordinal utility concept, marginal utility and interpersonal utility comparisons are meaningless. Utility simply is a numerical representation of preference and not a measure of happiness, well-being, or strength of preference.

Cardinal Utility

Cardinal utility, on the other hand, assigns meaning to utility differences which allow to capture, for example, diminishing marginal utility in riskless choices. Moreover, cardinal utility is frequently used to model an agent's risk attitude. The application of utility functions to risky decisions goes back to one of the first appearances of utility in a work by Bernoulli (1954, original 1738) who proposed expected utility as solution to the St. Petersburg paradox.

The ambiguous meaning of utility differences—as diminishing marginal utility and as formalization of risk attitudes—was already noted by Marshall (1997, original 1890). After von Neuman and Morgenstern (1944, 1947) extensively used the concept of utility to investigate decisions under risk, the distinction became even more relevant. Furthermore, cardinal utility is employed for inter-temporal evaluations, i.e. in all models where utility is discounted or aggregated over time, and for welfare considerations that transcend Pareto optimality.

Expected utility theory models cardinal utility from risky outcomes by von Neumann–Morgenstern utility functions. The same functions, however, do not directly apply to riskless choice analyzed in multi-attribute utility theory, for example. The ambiguity of the utility concept requires that cardinal utility functions are restricted to specific domains were they apply. Within this analysis, preferences over multi-issue goods in negotiations are most relevant—ordinal as well as cardinal utility functions will be of concern. Thereby, it will be pointed out if a utility function is cardinal. Furthermore, it is inevitable to refer to cardinal models of utility from risky outcomes as well. Most prominently, this will be the case in Sections 2.1.2, 2.2.1, and 2.2.3.

Rational Choice

Neoclassical models usually assume that agents are rational. What is rationality in this sense? When is a choice made rationally?

First of all, rationality is a property of patterns of choices, not singular choices—a single choice can never be irrational in itself. In order to judge rationality, one has to look at the change of choices when the available alternatives change or one has to know the agent's preferences. Rational choice is defined as a choice that can be explained by a preference relation.

Definition 2.4 (Rational Choice). *Given the preference relation $\succ$ and a choice set X, $x' \in X$ is a rational choice if there exists no $x'' \in X$ such that $x'' \succ x'$.*

A preference relation can in turn be represented by utility as presented above. The term rational choice thus specifies a choice that can be explained by an asymmetric and negatively transitive preference relation which by definition is a utility maximizing choice. Oftentimes utility functions solely capture an agent's self-interest, i.e. her self-regarding preferences. Self-interest is, however, not formally implied. It rather is generally assumed axiomatically. Some models like the one presented by Fehr and Schmidt (1999) or Bolton and Ockenfels (2000), for example, include fairness and equity considerations in standard utility models.

A normative theory of choice defines which choices can be classified as rational. It advises what to choose, for example, by advising to maximize a utility function. On the contrary, a descriptive theory portrays how agents behave. A description in this sense is not an explanation. It does not require the agents to deliberately carry out the considerations and calculations inherent in the theory. A description rather says that behavior looks *as if* the agent did follow the abstract mathematics. Allingham (2002, p. 10) illustrates this by the following analogy:

> 'A good description of the way a tree grows is obtained by assuming that it develops leaves in a way which maximizes the area exposed to the sun. But not even a tree-hugger would seriously suggest that the tree does this deliberately.'

A descriptive theory classifies a pattern of choices, i.e. a set of several choices, as irrational if no preference ordering exists which could explain all choices within the pattern. A degenerated pattern consisting of only one choice can always be explained by a preference ordering—consequently it cannot be irrational in itself.

Summary

The above presentation exemplifies a standard microeconomic model of preference, utility, and rational choice.

The most important issues to bear in mind for the subsequent analysis are the following: Microeconomic models of preference are concerned with alternatives an agent can choose from. If the alternatives are characterized by several issues, it is assumed that the agent can trade these issues off against each other (cf. negative transitivity). Preferences are formalized via preference relations and utility functions. These formalizations are, like any model, simplified models of reality. As such they are tractable and powerful theoretical tools. However, they necessarily leave out many real life influences on decision-making like cognitive processes and the context that leads to a decision.

2.1.2 Limited Descriptive Validity of Rational Choice

The above concept of preferences and individual choice is a widely used device for economists—it is applied to both prescriptive and descriptive models. However, there are several empirical studies which show limited descriptive validity of this model; standard economic theory does not fit the facts (Starmer, 2000). Nor do economic theorists necessarily want their models to fit the facts, this depends on the relative weighting of coherence and congruence.

Two of the most prominent examples of the empirical shortcomings of rational choice models are *framing effects* and *response mode effects*. Both examples challenge the assumption that preferences are well-defined prior to a choice situation. And these challenges to assumptions are a major contribution of experimental economics to the discipline at large. Rubinstein (2001, p. 619), for example, points this out from the theoretical perspective by stating that 'In any case, experimental economics should relate to the plausibility of assumptions we make on human reasoning rather than trying to accurately predict human behavior.'

Framing Effects

Neoclassical models on individual choice have two implicit assumptions which are rarely discussed in economic models (Starmer, 2000):

- *Description invariance* presumes preferences to be independent of the way outcomes are described.
- *Procedure invariance* implies that preferences over outcomes are independent of the method used to elicit them.

Description invariance neglects the influence of human perception on decisions, i.e. the formulation of a choice situation does not affect the choice (Arrow, 1982). Tversky and Kahneman (1981), among others, empirically demonstrate a violation of this implicit assumption. They frame one and the same decision in different ways before presenting it to subjects in an experiment and observe different decisions depending on the framing. The term *decision frame* thereby denotes an agents's perception of available actions, contingencies of the actions, and outcomes.

Asian Disease Framing Example

Oftentimes, a given problem can be framed in different ways as a famous and frequently cited example by Tversky and Kahneman (1981, p. 453) shows. Two groups of subjects face two different problem definitions. Both problems start with a common introduction:

> 'Imagine that the U.S. is preparing for the outbreak of an unusual Asian disease, which is expected to kill 600 people. Two alternative programs to combat the disease have been proposed. Assume that the exact scientific estimate of the consequences of the programs are as follows:'

Subsequent to the common introduction, group I is presented the following options:

> 'If program A is adopted, 200 people will be saved.
> If program B is adopted, there is a 1/3 probability that 600 people will be saved, and a 2/3 probability that no people will be saved.'

The expected number of people to be saved and people to die is equal for both programs. In the experiment by Tversky and Kahneman, the majority of subjects in this group were risk-averse: 72% of 152 subjects chose the certain outcome of program A over the risky program B.

Group II faced a different frame of the fundamentally same options:

> 'If program C is adopted, 400 people will die.
> If program D is adopted, there is a 1/3 probability that nobody will die, and a 2/3 probability that 600 people will die.'

Again, both programs are equivalent with respect to the expected number of people to be saved. Moreover, program C is the same as program A and program D is just another framing of program B. Consequently, description invariance along with random assignment of subjects to groups leads to the prediction that most subjects in group II should choose program C over program D. However, 78% of 155 subjects in group II preferred program D over C. When faced with a potential loss of lives, subjects tended to be risk taking.

Conflict with Rational Choice Models

The outlined data along with several studies on other choice situations as well as non-student subject pools reveal a common pattern: people tend to be risk-averse when choosing among gains while they are likely to be risk taking when choosing among losses even when the alternatives evaluated are objectively equivalent. This framing effect contradicts the rational choice assumption of asymmetric preferences (cf. Definition 2.1).

One behavioral explanation for the framing effect is that different presentations of a decision task invoke different mental processes which are based on potentially different orderings of outcomes (Slovic, 1995; Bettman, Luce, and Payne, 1998). Consequently, a single preference relation is not sufficient to explain individual choice in different situations.

Response Mode Effects

As discussed before, rational choice models implicitly assume description invariance and procedure invariance. Description invariance and the asymmetry of strict preference are challenged by the framing of decision tasks as shown in the previous section. This section presents empirical evidence questioning procedure invariance—the multi-issue trade-offs implied by negative transitivity of strict preference are not unanimous and independent of the procedure used to elicit preferences.

The contingent weighting of issues in multi-issue choice was, among others, demonstrated by Tversky, Sattath, and Slovic (1988) in an experiment on different response modes in two-issue scenarios.

Response Modes

The subjects' decision task was to select a job candidate. More specifically subjects were instructed to assume they would be an executive of a company and would have to select one of two applicants for the position of a production engineer. The information given is the rating of both applicants with respect to two issues chosen by the experimenters: technical knowledge and human relations. According to the experiment's instructions, technical knowledge was more important than human relations. The specific trade-off across issues was left unspecified and had to be decided by subjects.

The first response mode used by Tversky et al. was *choice mode*. The subjects' instructions contained a table like Table 2.1. The numbers are the job candidates' ratings where 100 is superb and 40 is very weak.

Table 2.1. Alternative choices in a response mode experiment (cf. Tversky, Sattath, and Slovic, 1988, p. 372)

	Technical knowledge	Human relations
Candidate X	86	76
Candidate Y	78	91

The table was presented to 63 subjects. 65% of them chose candidate X, i.e. the candidate with the higher score on the more important attribute. Thus, candidate Y's advantage with respect to human relations could not make up the shortcoming in technical knowledge for most subjects in choice mode.

The second response mode used was *matching*. Subjects got the same table as in choice mode, except that one of the four ratings was missing. Then they were asked to fill in the missing value so that both candidates would be equally suitable for the job. Under the assumption of procedure invariance, one can infer a subject's choice from her response to the matching task. If, for example, the technical knowledge of candidate X would be unspecified, i.e. the upper left value in Table 2.1 is missing, a subject's response for filling this filed might be 80. The actual value of 86 is higher than the subject's response. Thus, one can infer that the subject would choose candidate X in choice mode. If the subject would answer 90, for example, the inferred choice would be candidate Y.

Matching mode was played by four groups of about 60 subjects each. The groups differed with respect to which of the four numbers in Table 2.1 was unspecified. The differences in responses across these four groups are not significant. Averaged over all groups, 34% of subjects in matching mode preferred candidate X over Y compared to 65% in choice mode; the difference is significant.

Conflict with Rational Choice Models

Response mode effects are in conflict with an inherent assumption in rational choice models, namely procedure invariance. Choice and matching procedures are strategically equivalent. Consequently, random assignment of subjects to response modes should yield no systematic differences. The percentage of subjects favoring candidate X over Y should be the same for both response modes.

Procedure invariance is challenged by the data presented. Similar results were obtained by Tversky et al. in experiments on job candidates rated with respect to different issues, with environmental decisions to make, the choice among health care plans, and with decisions on future monetary payments. Comparable response mode effects were already demonstrated by Slovic (1975) and the pattern of observed discrepancies is persistent: the more important issue has a higher impact in choice than in matching. Tversky et al. present this as their *prominence hypothesis*.

The reasoning is that in choice mode, the more important issue, i.e. technical knowledge in the above example, looms larger and most subjects favor candidate X. In matching mode on the contrary, the relatively large advantage candidate Y has on the less important issue becomes more crucial and most subjects favor candidate Y.

The behavioral explanation of observed response mode effects is that different elicitation procedures focus the agents' attention on different aspects of the alternatives. Thus agents use different procedures, i.e. heuristics, to process the information and make a decision. These different heuristics result in seemingly inconsistent responses. Response mode effects are revisited in Section 2.3.2 with respect to other response modes and a more detailed discussion of preference construction.

Tversky et al. point out that procedure invariance is likely to hold when agents have well-defined preferences, for example, because they have sufficient experience with the domain and the response mode. Furthermore, procedure invariance might even hold if there is no master-list of preferences, but agents use a predefined algorithm to compute preferences. However, empirical evidence suggests that there are situations in which neither of these suppositions is satisfied and preferences depend on the elicitation procedure. This challenges the assumption of fixed multi-issue trade-offs implied by negative transitivity of strict preference (cf. Definition 2.2).

Summary

This section presented two widely accepted behavioral patterns—framing and response mode effects—which cannot be explained by microeconomic models of preferences and rational choice. Over the last decades, several behavioral extensions of the basic model have been proposed to account for the accumulated empirical evidence. The most prominent of these is prospect theory—it is covered in the following.

2.2 Prospect Theory

Prospect theory aims at overcoming descriptive limitations of the neo-classical model: it puts more emphasize on congruence with reality than on coherence of the model. The first version of prospect theory was proposed in a seminal paper by Kahneman and Tversky (1979) as alternative to expected utility theory; it is concerned with decisions under risk and outlined in Section 2.2.1. Twelve years after presenting prospect theory for decisions under risk, Tversky and Kahneman (1991) transferred the basic concepts to multi-issue choices in risk-less domains; this later development—which is vital for the subsequent analysis of multi-issue negotiations—is summarized in Section 2.2.2.

2.2.1 Decision Under Risk

The original model by Kahneman and Tversky (1979) deals with reference dependence, loss aversion, and probability weighting in decision-making over lotteries. It is a procedural theory in the sense that it describes decision-making as mental process.

Editing Decision Problems

In the first version of prospect theory, choices are a two-phase process: at first, a decision problem is edited, i.e. the agent reformulates it, and then the decision is based on this edited version of the problem in the second phase of decision-making.

Editing a decision problem is done by applying decision heuristics—heuristics are simple rules of thumb. The advantage of heuristic decision-making is a reduced mental effort. On the other hand, heuristics do not necessarily come up with an accurate result and can lead to systematic cognitive biases in some settings. The selection of a heuristic—whether consciously or unconsciously—is an accuracy-effort trade-off (Payne, Bettman, and Johnson, 1993; Smith and Walker, 2000).

One of the major editing heuristics proposed in prospect theory is coding outcomes as gains or losses relative to a reference point. An objectively given outcome of, for example, getting € 100 is internally perceived as gain of money or loss of money by an agent depending on what the agent's reference point is. The agent's reference might be that she expected to get € 200. In this case getting € 100 is a loss. Typically, the reference point is the status quo, but it can as well be influenced by the problem formulation and the agent's aspiration. The origin of reference points will be discussed in Chapter 3 and is focussed in the experiments reported in Chapters 4 and 5 Coding outcomes as gains or losses is a central argument in explaining subjects' responses in the Asian disease framing example presented in Section 2.1.2.

Other editing heuristics deal with simplifying alternative lotteries to choose: Combining probabilities associated with identical outcomes, rounding probabilities, rounding outcomes, eliminating elements common to all lotteries under consideration, and eliminating stochastically dominated options are some examples of decision heuristics that might be applied in the editing phase of decision-making.

It is important to note that prospect theory allows for this heuristics to be applied but it does not assure that they are applied by every agent and for every task. Which specific heuristics are applied is highly dependent on three factors: (1) the agent with her cognitive ability and her prior knowledge, (2) the decision problem with the context and task variables, and (3) the social context including, for example, group membership and the accountability for a decision (Payne, Bettman, and Johnson, 1993).

The editing phase is a major component of prospect theory in its first version (Kahneman and Tversky, 1979). Later on Tversky and Kahneman (1992) presented cumulative prospect theory, mainly to integrate the rank-dependent transformation of cumulative rather than individual probabilities that had been put forward by several authors (Weymark, 1981; Quiggin, 1982; Yaari, 1987; Schmeidler, 1989). Cumulative prospect theory still mentions the editing phase but there is no formal representation any more. The discussion of the editing phase is sourced out to Tversky and Kahneman (1986). Hence the procedural aspect of the theory is mainly abandoned. The editing phase is made obsolete as cumulative prospect theory allows the decision weighting function to be different for gains and losses (Starmer, 2000).

Choice Among Edited Prospects

Once the alternative lotteries, i.e. the prospects in the terminology of prospect theory, are edited, choice among them is determined in the second phase of decision-making. In the first version of the theory, the second phase was characterized by three properties: *loss aversion, reference dependence*, and *probability weighting*. In cumulative prospect theory, Tversky and Kahneman (1992) reduced this to two underlying more general properties: loss aversion and *diminishing sensitivity*. Diminishing sensitivity thereby explains the curvature of the value function and the probability weighting function.

Reference Dependence

Outcomes are not evaluated in absolute terms but as losses or gains relative to a reference point. Reference dependence allows to handle losses and gains quite differently; a property necessary to explain the Asian disease framing example. Reference dependence is formalized by a value function.[9] Figure 2.3 sketches a stylized value function. Let the function be denoted by $u(x)$, the first derivative by $u'(x)$ and the second derivative $u''(x)$. The reference point is $x = 0$. The function is generally assumed to be

- concave for gains ($u''(x) \leq 0$ for $x > 0$),
- convex for losses ($u''(x) \geq 0$ for $x < 0$),
- kinked at the reference point (thus u is not differentiable at $x = 0$), and
- steeper in the domain of losses ($\forall\, x > 0 : u'(x) < u'(-x)$).

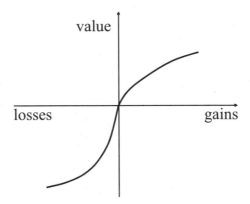

Fig. 2.3. Stylized value function in prospect theory

A value function like this implies risk taking behavior for losses and risk-averse behavior for gains; exactly the pattern reported by Tversky and Kahneman (1981) for the Asian disease framing example.

Loss Aversion

Loss aversion states that the impact of a difference in the domain of losses is experienced stronger by an agent than an equally sized difference in the domain of gains. Loss aversion is implied by the value function being steeper for losses than for gains.

---

[9] Kahneman and Tversky (1979) avoided the heavily used term utility function and used value function instead. The same terminology is applied here.

Figure 2.4 shows the effect of adapting a reference point. Initially, the reference point might be at $x = 0$. The analogy in a famous experiment by Kahneman, Knetsch, and Thaler (1990) is a subject possessing zero coffee mugs. In this situation, the subject's value for owning one coffee mug, i.e. $x = x'$, is given by $u(x')$.

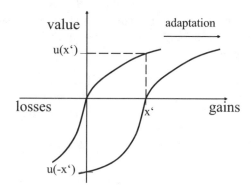

Fig. 2.4. Adaptation of reference points and loss aversion (cf. Strahilevitz and Loewenstein, 1998, Fig. 1a)

In their experiment, Kahneman et al. endowed subjects with coffee mugs and expected that this would shift the reference point and consequently the value function. The hypothesis here is that the status quo serves as reference point. A subject that owns a coffee mug and has a reference point at the status quo x', has a valuation of $u(-x')$ for loosing the mug. For value functions $u(x') < -u(-x')$ is assumed. Thus, loosing an object is perceived stronger than winning the same object—in prospect theory, agents are loss-averse.

Probability-Weighting

According to prospect theory, probabilities are not used objectively but distorted subjectively. Subjective probabilities are then used to calculate the expected value of alternative lotteries. The first version of prospect theory imposes some assumptions on a probability weighting function π; it is assumed to weigh relatively large probabilities less than they objectively are and to weigh relatively small probabilities more than they are. Furthermore, the function is monotonically increasing, its range is limited to $[0, 1]$, and the function is not defined for probabilities close to zero or unity. The latter property was abandoned later on.

In cumulative prospect theory, Tversky and Kahneman (1992) adopted the by then widely used inverted s-shaped function as probability weighting function. Figure 2.5 plots a stylized function with inverted s-shape.

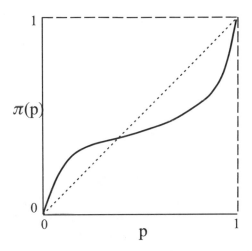

Fig. 2.5. Stylized probability weighting function in prospect theory

Cumulative prospect theory combines a s-shaped reference-dependent value function with an inverted s-shaped probability weighting function. Taken together, these functions imply a fourfold pattern of risk attitudes: they imply risk taking behavior for small-probability gains and large-probability losses and they imply risk aversion for small-probability losses and large-probability gains.

A lottery with an expected payoff of zero and a small-probability loss is, for example, the following: win € 1 with 95% chance and loose € 19 with a 5% chance. A decision-maker then could have the choice to either play the lottery—this is risk taking behavior—, or to refrain from playing—this is risk-averse behavior. If the small probability of 5% is overestimated and the large probability of 95% is underestimated, the subjectively expected payoff becomes negative. This is even strengthened if the value function enters the evaluation. Loosing money is perceived stronger than winning money. Thus, the expected utility of the lottery is negative and a utility maximizing decision-maker would choose not to play the lottery. This is the risk-averse behavior for small-probability losses implied by cumulative prospect theory. The other three patterns of risk attitudes can be demonstrated by similar examples.

Diminishing Sensitivity

In cumulative prospect theory, the curvature of the value function and the probability weighting function are interpreted as implications of the more general property diminishing sensitivity (Tversky and Kahneman, 1992). Diminishing sensitivity says that the impact of a marginal change diminishes as one moves farther away from the reference point. The functions sketched in Figures 2.3 to 2.5 can thereby be explained.

The value function intersects the abscissa at its reference point and it is monotonically increasing. Diminishing sensitivity calls for the function to be flatter at the far right of the reference point than close to the reference point. Hence the function must be concave for gains. Analogously, diminishing sensitivity results in convexity for losses.

The inverted s-shaped probability weighting function can be explained by diminishing sensitivity as well. A function with diminishing sensitivity is steepest close to a reference point. If the probabilities of zero and unity are two natural reference points, then the function must be steep at its lower and upper end and relatively flat in an intermediate range—this is an inverted s-shape. Hence the principle of diminishing sensitivity explains the form of the value function as well as the form of the probability weighting function (Starmer, 2000).

Calibration of Prospect Theory

Up to now, value function and probability weighting function were introduced qualitatively. Stylized facts of these functions can be empirically observed. On the contrary it is impossible to observe the specific functional form. Furthermore, it might very well be that there exist no such functions. Just as the neoclassical model outlined before, prospect theory is a simplified view of the world. Its an inherently descriptive rather than a normative theory and assumes that agents act *as if* guided by a value function and a probability weighting function.

Calibration of the Value Function

Several researchers have proposed specific formalizations that capture the stylized facts observed. The parameters of these functions can be estimated with experimental data. Most prominently, Tversky and Kahneman (1992) proposed the following two-part power function as value function:

$$u(x) = \begin{cases} x^\alpha & \text{if } x \geq 0, \\ -\lambda(-x)^\beta & \text{otherwise} \end{cases}$$

The parameters α and β determine the curvature for gains and losses, respectively. Furthermore, λ is the degree of loss aversion. Tversky and Kahneman estimate the parameters as $\alpha = \beta = 0.88$ and $\lambda = 2.25$.

Calibration of the Probability Weighting Function

Different functions have been proposed for probability weighting. Tversky and Kahneman (1992) suggest

$$\pi(p) = \frac{p^\delta}{(p^\delta + (1-p)^\delta)^{1/\delta}}$$

which leads to an inverted s-shaped function for $0 < \delta < 1$. Reducing δ lowers the crossover point of the inverted s and the diagonal line, i.e. the only probability except zero and unity that is perceived objectively. Tversky and Kahneman estimate the parameter δ for gains and losses separately. They find $\delta = 0.61$ for gains and $\delta = 0.69$ for losses and conclude that the probability weighting function is essentially the same for gains and losses. Figure 2.6 plots the weighting functions for gains and losses; they are labeled 'T&K 92 (gains)' and 'T&K 92 (losses)', respectively.

Prelec (1998), on the other hand, proposes the alternative functional form

$$\pi(p) = exp(-(-\ln p)^\alpha)$$

with $0 < \alpha < 1$ as probability weighting function. It generates an inverted s-shape as well. As α approaches zero, the function becomes linear; as it approaches unity, the function essentially becomes a step function. Prelec estimates the parameter to be $\alpha = 0.65$. The respective function is plotted in Figure 2.6. It is almost identical to the function Tversky and Kahneman found for losses.

2.2.2 Riskless Multi-Issue Choice

Reference dependence, loss aversion, and diminishing sensitivity carry over from risky choices to riskless multi-issue choices (Tversky and Kahneman, 1991). For this, the basic neoclassical model introduced in Section 2.1.1 has to be extended to account for reference points. With multiple issues, choosing among alternatives oftentimes requires trading off one issue against another. Under the assumption of negatively transitive preferences this is always possible. However, a reference point can influence the marginal rate of substitution in these trade-offs.

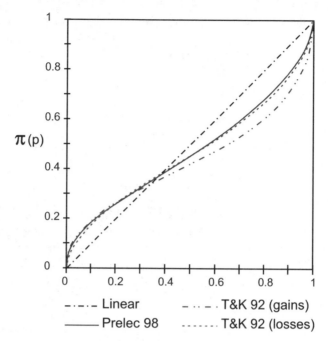

Fig. 2.6. Estimated probability weighting functions in prospect theory (cf. Prelec, 1998, Fig. 1)

Theoretical Framework

Let X be a finite set of alternatives the agent can choose from, just like in Section 2.1.1. Furthermore, let each alternative be characterized according to several issues, also known as attributes or dimensions. Each $x \in X$ is interpreted as a vector of several issues $x = \langle x_1, x_2, \cdots, x_n \rangle$ with integer $n \geq 2$ and $x_i \geq 0 \ \forall \ i \in \{1, 2, \cdots, n\}$.[10]

A *reference structure* is a family of indexed preference orderings $\succ_r$ where $x \succ_r y$ stands for alternative x being strictly preferred to y evaluated from reference state $r \in X$ (Tversky and Kahneman, 1991). The relation $\succ_r$ is assumed to be asymmetric and negatively transitive (cf. Definitions 2.1 and 2.2). Weak preference $\succeq_r$, indifference $\sim_r$, and a utility representation $u_r(x)$ follow analogously to the derivation in Section 2.1.1.

A single, fix reference point can easily be assumed in a traditional neoclassical model as the model itself is agnostic to whether alternatives

---

[10] The notation '$\langle \cdots \rangle$' is used throughout the study to denote vectors that stand for choice alternatives. Usually, this will be points in the agreement space of a negotiation, i.e. offers, agreements, and reference points.

are perceived as gains or losses. The neoclassical model does, however, not allow for the change of a reference point. Hence, it is a special case of the prospect theoretical model where the reference point is constant.

Introducing reference dependence raises two questions: (1) how does an existing reference point influence choice and (2) where does the reference point come from in the first place? Tversky and Kahneman (1991) outline the answer to the first question, i.e. the effect of reference dependence on choice, in great detail. Their argumentation is the foundation for the remainder of this section on riskless multi-issue choice. They are, however, not concerned with the emergence of reference points (p. 1046). The emergence of reference points, especially in bilateral negotiations, is addressed in the following Section 3.2.

Loss Aversion

Losses are perceived stronger than gains. Hence the slope of indifference curves changes depending on the reference point. The following definition with its several variables and indices is exemplified by Figure 2.7 for the two-issue case; thereby issues i and j from the definition correspond to issues 1 and 2 in the figure.

Definition 2.5 (Loss aversions). *A reference structure satisfies loss aversion if* $\forall\, x, y, r, s \in X$, $i, j \in \{1, 2, \cdots, n\}$, *and* $i \neq j$ *with* $x_i \geq r_i > s_i = y_i$, $y_j > x_j$, *and* $r_j = s_j$ *the following condition holds:* $x \sim_s y \Longrightarrow x \succ_r y$.

Note that $\geq, >$, and $=$ refer to the numerical values on single issues whereas $\succ$ and $\sim$ are relations of multi-issue alternatives. The agent is assumed to prefer more to less on each issue. Figure 2.7 exemplifies the definition: Alternatives x and y are the ones to choose from, points r, s, and t are reference points[11]

There is no dominance relation between x and y. Thus, the agent has to trade the differences in issues 1 and 2 off against each other. Assume that the agent is indifferent when evaluating the two alternatives from reference point s, i.e. $x \sim_s y$ as required in the definition. Both alternatives lie on the same indifference curve labeled u_s.

What changes if x and y are evaluated from point r? The difference in issue 2 is to the advantage of alternative y ($y_2 > x_2$). It is unaffected by the change in the reference point, as $r_2 = s_2$ holds due to the condition $r_j = s_j$ in the definition. On issue 1, the difference of x and y favors x ($x_1 > y_1$). This difference is affected by switching from s

---

[11] Point t will be used subsequently for illustrating diminishing sensitivity.

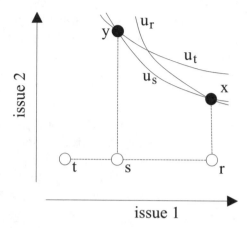

Fig. 2.7. Reference dependence in multi-issue choice (cf. Tversky and Kahneman, 1991, Fig. III)

to r. Seen from s, both y and x offer a gain on issue 1, even if it is zero for y. From r, x_1 still is a gain, while y_1 is perceived as a loss as $x_1 \geq r_1 > s_1 = y_1$ holds.

The difference $x_1 - y_1$ is independent of the reference point. However, from s it is evaluated as the difference of two gains and from r as the difference of a gain and a loss. Losses loom larger than gains. Hence, the difference becomes more pronounced in the overall decision when evaluated from r. From s, alternatives x and y appeared equally good. From r, the difference in issue 1 appears more severe and cannot be compensated by the advantage alternative y has on issue 2. Thus, the agent strictly prefers x over y when evaluated from r ($x \succ_r y$); the marginal rate of substitution across issues changes as the reference point changes.

Diminishing Sensitivity

Marginal utility decreases with increasing distance from the reference point. Hence, in Figure 2.7 the evaluation of the difference of alternatives x and y on issue 1 will depend on whether it is seen from reference point s or t.

Definition 2.6 (Diminishing sensitivity). *A reference structure satisfies diminishing sensitivity if $\forall\, x, y, s, t \in X$, $i, j \in \{1, 2, \cdots, n\}$, and $i \neq j$ with $x_i \geq y_i$, $y_j > x_j$, $s_j = t_j$, and either $y_i \geq s_i \geq t_i$ or $t_i \geq s_i \geq x_i$ the following condition holds: $y \sim_s x \Rightarrow y \succ_t x$.*

Again, the definition is illustrated by Figure 2.7 by assuming $i = 1$ and $j = 2$. On issue 1, x is as least as good as y ($x_1 \geq y_1$) while y

has an advantage on issue 2 $(y_2 > x_2)$. The two differences have to be traded off against each other. From reference point s, they just equal out $(y \sim_s x)$.

What changes if x and y are evaluated from point t? For answering this, the relative location of s and t has to be defined. First of all, they are the same on issue 2 $(s_2 = t_2)$. Furthermore, either y_1 and x_1 are both perceived as gains $(y_1 \geq s_1 \geq t_1$; the option chosen in the figure) or both as losses $(t_1 \geq s_1 \geq x_1)$ and the distance on issue 1 from t to any of the alternatives x or y is greater than from reference point s.

Alternative x is better on issue 1, alternative y on issue 2. From reference point s the agent is indifferent $(y \sim_s x)$. When changing the reference point to t, the difference on issue 2 is unaffected. The difference on issue 1, however, is affected. It now is farther away from the reference point. The marginal effect of a difference diminishes with distance from the reference point. Hence, $x_1 - y_1$ is perceived less strongly and cannot anymore offset the advantage y has on issue 2—the agent prefers alternative y from reference point t, i.e. $y \succ_t x$.

An indifference curve represents a set of alternatives among which the respective agent is indifferent. In rational choice models indifference curves never intersect. If they would do, the agent would be indifferent among all alternatives on both curves and they would essentially reduce to a single indifference curve. This is due to the transitivity of the indifference relation in neoclassical modeling (cf. Sec. 2.1.1). In prospect theory, indifference curves drawn with respect to different reference points can intersect. This is exemplified in Figure 2.7. The transitivity of $\sim$ carries over to each single one of $\sim_r$, $\sim_s$, and $\sim_t$ independently, as long as the reference point is fix. It is, however, incorrect to build chains of transitive reasoning over different reference points like $x \sim_s y \sim_r z \Longrightarrow x \sim z$.

In the example (Figure 2.7), there is a strict interrelation of reference points and alternatives. The reference points are all the same on issue 2 and two of them coincide with the alternatives on issue 1. This is, however, not a necessary condition for the outlined concepts—namely loss aversion and diminishing sensitivity—as the definitions show. The implications can be extended to situations where the reference points do not coincide with the alternatives on any issue.

Implications and Evidence

Different outcomes are compared via their differences on single issues. The relative weight of these differences depends on the location of the reference point. More specifically, it depends on the distance to the reference point (diminishing sensitivity) and whether the difference is perceived as a gain or a loss (loss aversion). Hence, an agent's preferences over multi-issue alternatives depend on her reference point.

The origin of the reference point has not yet been discussed. However, there is evidence that expectations and market processes play a vital role; see for example Köszegi and Rabin (2006) for a theoretical model, Heyman, Orhun, and Ariely (2004) for empirical evidence, and Section 3.2 for a more thorough discussion of the matter.

In the neoclassical model, preferences are captured without reference to the current state or any other reference point. In this model, the Coase theorem assures that initial property rights do not influence the final allocation of goods—the efficient allocation will be reached by market transactions, like bilateral exchange, if one neglects transaction costs (Coase, 1960). With reference dependence, this does not hold any more. If market processes and expectations influence reference points, then they influence agents' preferences and their willingness to trade goods. A market process may become an irreversible process and the final allocation of goods might not be efficient with respect to the agents' a-priori preferences, i.e. their preferences before the process started.

Empirical Evidence

Tversky and Kahneman (1991) designed several experiments to test the implications of prospect theory for riskless multi-issue choice. In one of their experiments, subjects were promised a gift and were then required to choose among three options: (1) keep the original gift, (2) exchange it for two vouchers for free dinners (option x), or (3) exchange it for a voucher for one large and two small professional photo portraits (option y).

For half of the subjects, the so called dinner group, the original gift was one voucher for a free dinner and a calendar. For the other half, the so called photo group, it was a voucher for one large professional photo portrait and a calendar. At the end, the gifts were assigned at random to some subjects.

Table 2.2 outlines the setting: the columns give the four issues at hand and the rows represent the alternatives. Subjects were assigned to either of the alternatives dinner or photo and could then choose to keep this alternative or to change to option x or y.

The assumption by Tversky and Kahneman was that promising the gifts influences the subjects' reference points and hence their decision. They assumed that the original gift would be adopted as reference point. Only 10 out of 90 subjects kept their original gift. The remaining

Table 2.2. Alternative choices in an improvements versus trade-offs experiment

	issues			
	1	2	3	4
	dinner voucher	large photo	small photo	monthly calendar
dinner	1	0	0	1
photo	0	1	0	1
option x	2	0	0	0
option y	0	1	2	0

80 subjects exchanged it for either option x or y. Loss aversion predicts which exchanges should occur.

A subject in the dinner group for sure looses on issue 4 if she chooses either x or y. This difference does not give a prediction on the specific choice. If the subject chooses x, however, she wins on issue 1 and there is no change on issues 2 or 3. If she chooses y, she perceives a loss on issue 1 and gains on issues 2 and 3. Loss aversion implies that subjects in the dinner group will tend to favor option x over y due to the loss on issue 1 that would be caused by choosing y. In fact, 81% of subjects in the dinner group that changed their gift chose option x.

The reasoning for the photo group is analogous. Options x and y are equal on issue 4. Choosing x results in a loss on issue 2. Choosing y does not cause a loss on any of the issues 1, 2, and 3—it is an improvement on these issues and there is no trade-off necessary. Hence, the prediction is that loss-averse subjects tend to choose option y. In fact, 52% of subjects in the photo group that changed their gift chose option y. With random assignment of subjects to groups, there should be no difference in the proportion favoring one or the other option. The observed difference of 81% in the one and 52% in the other group is, however, statistically significant and thus supports the relevance of loss aversion for multi-issue choice (Tversky and Kahneman, 1991).

Estimating Loss Aversion

There is not one general degree of loss aversion. Loss aversion depends (at least) on the individual, the context of the choice situation, the elicitation procedure, and the issue in which the gain or loss occurs (Tversky, Sattath, and Slovic, 1988). In a study on health risks, for example, Viscusi, Magat, and Huber (1987) find that loss aversion seems to be stronger for safety than for money. Based on data by Kahneman, Knetsch, and Thaler (1990), Tversky and Kahneman (1991) estimate the degree of riskless multi-issue loss aversion for consumer goods, specifically for coffee mugs, to be about 2, i.e. losses loom twice as large as gains.[12]

Interestingly, this riskless multi-issue loss aversion seems to be about the same as the estimated loss aversion in risky choices (cf. Sec. 2.2.1). This should, however, be interpreted with caution as both measures depend on the specific issues and situations as well as on the functional forms assumed and data sets used for estimation. Furthermore, concepts and especially numerical estimates from the theory of risky decisions cannot directly be applied for riskless choices, as outlined in Section 2.1.1.

2.2.3 Alternative Non-Neoclassical Utility Theories

Neoclassical utility theory is a simple and powerful model for individual decision-making. It is, however, not fully capable of explaining empirically observed behavior. Non-neoclassical utility theories try to overcome this limited descriptive validity. Basically, there are two classes of such theories: procedural theories and utility optimization theories (Starmer, 2000). Prospect theory is a non-neoclassical utility theory—in its first version it clearly is procedural; cumulative prospect theory, on the other hand, puts less emphasis on the procedure and more on utility optimization. This section briefly outlines alternatives to prospect theory.

Procedural Theories

In procedural theories decision-makers are assumed to base their choices on heuristics. Agents choose a specific decision rule depending on a given context. Thereby, the quality of the decision might be traded off against unusual alternative objectives like (mental) costs for information processing—Payne, Bettman, and Johnson (1993) talk about accuracy-effort trade-offs. Besides prospect theory, other procedural theories are the general model on procedural decision-making presented

---

[12] Franciosi et al. (1996) replicated some settings of Kahneman, Knetsch, and Thaler (1990) and reformulated the instructions in a way that all references to 'buying', 'selling', and 'prices' were removed. This replication confirmed the results of Kahneman et al. qualitatively although less pronounced.

by Payne, Bettman, and Johnson (1993)[13] and the theory on similarity-based preferences in decision-making put forward by Rubinstein (1988).

Utility Optimization Theories

Utility optimization theories are closer to neoclassical theory than procedural theories are. They do not model the psychological process of decision-makers but retain the assumption of preference maximization and behavior *as if* optimizing an underlying utility function. There are mainly two conventional ways to extend neoclassical expected utility theory: either subjective weights are assigned to outcomes which are then combined with objective utilities, or objective probabilities are transformed to subjective probabilities which are used fore aggregating different outcomes. Starmer (2000) presents an excellent overview on both categories of extensions.

A widely recognized model with subjective evaluation of outcomes is *regret theory* introduced by Loomes and Sudgen (1982). Utility is derived from two sources: First of all there is a choiceless utility, i.e. a utility an agent would derive from an outcome if she experienced it without having chosen it. Furthermore, there is a regret and rejoice function—it is assumed that an agent regrets a choice if, later on, it turns out that another choice, like entering another lottery, would have been better. Analogously, the agent rejoices in having chosen the best alternative possible. Overall utility is a combination of choiceless utility and regret or rejoice. The utility experienced by an agent does not only depend on the final outcome, but as well on the agent's expectations. The expectations inherent in regret theory are closely related to the reference point in prospect theory—outcomes are not evaluated in absolute terms but relative to some point: either an expectation, or a reference. In fact, in Section 3.2.2 it will be argued that expectations influence reference points.

Several utility optimization theories, like the *theory of anticipated utility* by Quiggin (1982), incorporate inverted s-shaped decision-weighting functions. This transformation of objective to subjective probabilities is a pivotal element of prospect theory as well.

Quality of the Theories

The plethora of traditional and behavioral economic theories for individual decision-making naturally leads to questioning the quality

---

[13] The model by Payne et al. is discussed in Section 2.3.1.

of these theories—which theory is best? Answering this question inevitably requires trading off the relative importance of a theory's characteristics like generality, manageability, tractability, congruence with reality, and predictive accuracy (Stigler, 1950). There is no universal answer to the question which theory is best.

A major benefit of neoclassical utility theory certainly is its analytical tractability. On the other hand, non-neoclassical theories are oftentimes advantageous with respect to predictive accuracy. Over the last decades, there has been a trend towards non-neoclassical theories (Starmer, 2000). However, these relatively new theories are still far from replacing neoclassical utility theory on a general basis; they are rather substitutes in limited domains. Prospect theory seems promising in domains where one has a clear idea of what the agents' reference points are and how changes in reference points are determined. The following chapters are organized around the question how changes of reference points are determined in negotiations if there are changes at all.

2.3 Preference Construction and Stabilization

Behavioral psychology and microeconomics rest upon different notions of preferences: economists usually assume preferences to exist and to be exogenously given or to be context-dependent but fixed (up to changes of the reference point) like in the behavioral model outlined in Section 2.2. On the contrary, the behavioral school of constructive preferences argues that preferences are constructed via information processing. This view is less abstract than utility optimization and intends to show greater correspondence to reality, i.e. the real working of a human brain, than economic models.

Preferences are not merely revealed but constructed by agents at the time a choice has to be made. This construction process is task and context dependent. Observed choices are more than the result of a lookup in a master list of preferences agents have in mind. Some authors even argue that there is no such thing as a fundamental value of an item to an agent. This perspective on preferences is persistently taken over the last decades, e.g. by Tversky, Sattath, and Slovic (1988), Payne, Bettman, and Johnson (1993), Bettman, Luce, and Payne (1998), Hoeffler and Ariely (1999), Slovic, Griffin, and Tversky (2002), and Ariely, Loewenstein, and Prelec (2006).

The literature on constructive preferences mainly bases on marketing studies and consumer choice theory. Oftentimes it is assumed that agents employ heuristics, i.e. simple rules of thumb, to restructure problems, process information, and finally make a choice.

In the theory of constructive choice processes decisions are entirely based on heuristics. Thus, this theory is one step further away from utility maximization models than prospect theory is. Prospect theory allows for heuristics in agents' perception of decision tasks. Editing an absolute outcome as relative gain or loss is one such perceptual heuristic. Other models of constructive preferences, like the accuracy-effort trade-off model described below (Payne, Bettman, and Johnson, 1993), focus more on how information is processed heuristically once it is perceived.

The effect of reference dependence is precisely modeled in prospect theory. The emergence of reference points on the other hand is hardly addressed. Some economists might be tempted to think that the origin of reference points is a psychological and not an economic issue (Starmer, 2000). However, the exact way reference points emerge in economic situations has a major influence on economic behavior in some domains. This is, for example, outlined by Heath, Huddart, and Lang (1999) in a study on employees exercising stock options. The authors show that the employees respond to stock price trends in a way that is inconsistent with purely maximizing monetary payoff but can be explained by reference-dependent evaluations.

Heuristics that influence decision-making—via reference points or otherwise—are an important factor in understanding economic agents and their decision-making in, for example, negotiations. The following subsection looks at how agents decide which heuristics to apply.

2.3.1 Adaptive Decision Making Processes

Many people do not process all relevant information when making a decision but rather apply heuristics which use information selectively. Oftentimes people have different heuristics to solve one and the same decision task. Selecting a specific heuristic thereby is context and task dependent—agents decide how to decide (Tversky, Sattath, and Slovic, 1988; Payne, Bettman, and Johnson, 1993).

Selective Information Processing

Agents decide how to decide because they are constrained in memory organization and retrieval as well as in information capacity and processing. An accurate decision requires thoroughly processing all available information. Even if this would be possible, it requires substantial mental effort. Thus, agents have to trade off the accuracy of a decision and the effort applied to finding it.

Bounded Rationality

Full rationality, an assumption inherent in many economic models, requires unlimited information processing and cognitive capabilities. The concept of bounded rationality breaks with this theoretically striking but descriptively inaccurate assumption. Bounded rationality denotes the rational principles that underlie non-optimizing behavior of real people (Selten, 2001).

Going back to Simon (1955, 1957), bounded rationality should not be mistaken for irrationality. On the contrary, in the face of constraint cognitive bounds it might be perfectly rational to process information selectively. Thus, according to Camerer (1995, p. 97), 'rule-following adaptive agents' would be a more appropriate labeling than bounded rationality. Resolving the ambiguity among accuracy and effort is not necessarily a mistake in human behavior—it is a form of intelligence which can be incorporated in models of decision-making.

Accuracy-Effort Trade-Offs

The model by Payne, Bettman, and Johnson (1993) is a full-fledged model on boundedly rational individual decision-making. The authors assume that cognitive effort is a scarce resource and that, given this limited information processing capacity, agents try to attain decision accuracy as well as limiting the cognitive effort devoted to solving a task.

The fact that processing information is perceived as effort has been pointed out by, for example, Gneezy and Rustichini (2000). In one of their experiments, students were required to answer questions from an IQ test in a given time. Different groups of students were offered different amounts of money for each correct answer. Students receiving a relatively low payoff perform worse than students receiving a relatively high payoff for each correct answer. Furthermore, an interesting point is that students not receiving money at all perform better than the low payment group. Gneezy and Rustichini conclude that there is intrinsic motivation for answering correctly if students do not receive a payment and a strong extrinsic motivation when offered a high payment. A low payment on the other hand crowds out the intrinsic motivation without being an sufficient extrinsic one. Hence, subjects are not willing to put mental effort in answering correctly and perform worse than the other groups.

The idea of costly information acquisition and processing calls for the information being easily accessible. However, the availability of information is not sufficient for it being processed. Russo (1977) studied the value of unit price information in supermarkets. His hypothesis was that displaying unit prices would ease the comparison across products and, consequently, consumers would tend to buy the products with lower unit prices. However, displaying the prices on separate shelf tags right beside every single product had just a weak effect on consumer behavior. Displaying the prices in a single ordered list, on the other hand, affected purchasing behavior much stronger. Thus, an accuracy-effort trade-off does not only require the availability of information but information needs to be processable. An easy presentation can reduce the mental effort associated with processing.

Decision-making is more than retrieving preferences from memory and choosing the most preferred alternative instantaneously. It rather is a mental procedure of information processing. Decision-making involves the sub-processes information acquisition, information evaluation, and expression of a decision. For each sub-process, an agent has several heuristics that can be applied, i.e. relatively simple rules how to solve a sub-problem (Payne, Bettman, and Johnson, 1993).

Decision Heuristics

Heuristics oftentimes solve problems in an efficient and satisfying way, but they do not guarantee a solution. Algorithms specifically tailored to a problem type, on the other hand, always solve problems of the respective type with certainty. They are, however, oftentimes more computationally demanding. The relation of algorithms and heuristics can be seen by an example on simple mathematics (Wessells, 1982, Ch. 9).

Example

Below, there are eight numbers to be summed up. Try to solve the addition in 30 seconds:

$$
\begin{array}{r}
85515 \\
14485 \\
3555 \\
6445 \\
85515 \\
14485 \\
3555 \\
+6445 \\
\hline
?
\end{array}
$$

The algorithm for solving the summation is taught in primary school: one starts with the last column and adds up the last digit of every number. The result is 40 in this example. The zero is put down in the last column, the four carries over to the second-last column, etc. The algorithm is easy to perform and guarantees a result. It is, however, relatively slow. Hardly anybody carries it out within 30 seconds for the above numbers.

A heuristic approach is to look for patterns in the numbers: the last four numbers equal the first four, for example. Furthermore, the first and second number add up to one hundred thousand and the third and fourth one to ten thousand. This way, the overall sum of two hundred and twenty thousand can be calculated faster than by using the algorithm. Thus, the heuristic uses an important resource, i.e. time, economically and nevertheless results in the correct answer. This is, however, not true in general as for other problem structures and numbers different heuristics or the algorithm can be superior—accuracy and effort have to be traded off.

Choice of a Heuristic

Agents decide how to decide; sometimes they do so consciously, sometimes unconsciously. The choice of a heuristic for information acquisition and evaluation is influenced by task and context factors. Task factors refer to the general structural characteristic of a decision task. Examples are the number of alternatives, the number of issues, time pressure, and agenda constraints. Context factors, on the other hand, depend on the specific alternatives under consideration in a single decision task. The similarity of alternatives and their overall attractiveness are two examples for context factors.

In multi-issue decision-making, maximizing a weighted additive utility function is often seen as a normative rule (Keeney and Raiffa, 1993).[14] However, many agents seem no to follow this 'gold standard' for decision-making but to apply heuristics that are cognitively less demanding.

For illustration of the following heuristics assume that an agent has to chose from an ordered list of alternatives $X = \langle A, B, C \rangle$ where each alternative is defined by three numerical issues x_1, x_2, and x_3. Further assume that the agent prefers a higher value on each issue, that issue

---

[14] Note that a multi-issue utility function can be expressed in weighted additive form if and only if mutual preferential independence of the issues holds. See Keeney and Raiffa (1993, Ch. 3) for the exact representation theorems which provide conditions for expressing utility functions in weighted additive form.

x_1 is the most important and issue x_3 the least important, and that the alternatives are given as follows: $A = \langle 5, 7, 1 \rangle$, $B = \langle 4, 3, 3 \rangle$, and $C = \langle 5, 6, 4 \rangle$. Finally, the agent has so called cut-off levels, i.e. minimum requirements, of 4, 3, and 2 units for the three respective issues.

Lexicographic Search

One heuristic is, for example, lexicographic search in analogy to lexicographic ordering of words. Issues are ordered by decreasing importance and the agent iterates over the list of issues. For each issue, the agent dismisses all alternatives that do not offer the highest value among all alternatives under consideration on the current issue. The search stops as soon as just a single alternative remains (Payne, Bettman, and Johnson, 1993).

According to lexicographic search, an agent chooses alternative A in the above example. First, issue x_1 is evaluated and alternative B is dismissed. A and C tie and remain in the set of considered alternatives. Next, issue x_2 is examined and C is dismissed. The only remaining alternative A is chosen.

Lexicographic search is a non-compensatory heuristic. It does not allow for the possibility that a failing in one issue is compensated by a more than satisfactory value on a different issue.

Satisficing Heuristic

Another simple non-compensatory heuristic is the satisficing heuristic. The agent sequentially searches the set of alternatives until she finds one alternative that satisfies the cut-off levels on all issues. This alternative is then chosen without considering the remaining ones. If none is chosen after iterating over all alternatives, either the cut-off levels are relaxed, or one alternative is chosen at random (Simon, 1955).

In the example, an agent following the satisficing heuristic chooses alternative B. A is processed first, as it appears first in the ordered list of alternatives. It is dismissed because it does not satisfy the cut-off level of 2 units on issue x_3. B is processed next. It is satisficing and thus chosen. C is not considered.

Elimination by Aspects

A third frequently applied decision heuristic is elimination by aspects. Again, the issues are ordered by decreasing importance and the agent iterates over them. Alternatives that do not satisfy a cut-off level on the current issue are eliminated from the set under consideration. The difference to the satisficing heuristic lies in the order of search: the satisficing heuristic processes information alternative-wise whereas elimination by aspects processes information issue-wise (Tversky, 1972).

All three alternatives in the example exceed the cut-off level on issues x_1 and x_2. On x_3, alternative A fails. For the final decision, the agent has to break the tie across B and C, for example, by noticing that C dominates B—hence, C is chosen.

Note that the alternative chosen by the satisficing heuristic remains satisfying when an agent uses elimination by aspects. Thus, elimination by aspects is as least as accurate as the satisficing heuristic. However, this accuracy is bought at the cost of a higher mental effort; in the example it required more comparisons than the satisficing heuristic.

Other heuristics deal, for example, with the majority of confirming dimensions of an alternative or with the frequency of good and bad features an alternative possesses. Both heuristics are compensatory.

Combined Heuristics

Oftentimes not a single heuristics is chosen, but a combination. Agents might, for example, eliminate poor alternatives via rough cut-off levels in a first step and then choose more carefully among the remaining alternatives by using a compensatory heuristic.

The choice of a specific heuristic or a combination thereof is task and context specific. Some general rules are presented in a meta-study by Bettman, Luce, and Payne (1998): (1) the more complex a task is in terms of the number of alternatives, the higher the probability of a non-compensatory heuristic; (2) an increase in the number of issues presented leads to more selective information processing; (3) under time pressure, agents become more selective as well and negative features are weighted more strongly; (4) agents are more likely to process information extensively and alternative-wise under negative inter-issue correlation, i.e. if the necessity of trade-offs across issues is more obvious; (5) emotion-laden decision tasks tend to induce more issue-wise information processing; etc.

Besides the heuristics briefly sketched here, there is an enormous amount of other heuristics persistently found in human decision-making. See Payne, Bettman, and Johnson (1993) and Gilovich, Griffin, and Kahneman (2002) for extensive compilations.

Adaptive Decision Making and Prospect Theory

The editing phase in prospect theory allows for agents applying heuristics to a decision task. 'The function of the editing phase is [...] to simplify subsequent evaluation and choice' (Kahneman and Tversky, 1979, p. 274). As mentioned before, the difference to the adaptive decision-making model by Payne, Bettman, and Johnson (1993) lies in the range of heuristics. In prospect theory, heuristics apply to the perception of a task and internal representation of the alternatives. In adaptive decision-making, heuristics are employed to make a decision based on this internal representation of alternatives. Furthermore, the adaptive decision-making model puts more emphasize on mental effort involved in decision-making and selective information processing.

Preference Stabilization

If it is the case that preferences are constructed and that this process is highly task and context specific, then how does it come that one observes consistent choices and stable revealed preferences in some domains and for some agents? The answer is that the construction process stabilizes the more often the same choice (or at least a similar choice) has been made. Agents gain experience with tasks and contexts and the outcome of the preference construction process stabilizes.

Some decision problems are simple memory retrieval tasks as they have been solved previously (Bettman, Luce, and Payne, 1998). If an agent is asked for his name, for example, there are numerous names to choose from. Nevertheless, the answer will likely be the same in a variety of situations as it is a simple memory retrieval exercise. This holds for relatively easy tasks and especially if the alternative choices are clearly distinguishable.

If, on the other hand, the task and the context have some novelty and complexity it is not that easy. It is likely that preferences are not readily available in memory. Instead, more effort has to be applied to construct them via heuristics. Depending on how closely the situation is related to previously encountered ones, the outcome of this construction process is more or less determined.

The model of discovered preferences is in between simple memory retrieval and full construction (Plott, 1996). It is assumed that agents have a well-defined set of preferences without being fully aware of it. By thought and experience agents can discover and then reveal their preferences.

Note that in this outline the focus is on preference construction and stabilization in a rather short run and for a single agent. A related point from a global long-term perspective is the evolution of preferences over generations of agents through genetic inheritance and cultural learning.

See Bowles, Choi, and Hopfensitz (2003) and Bowles (2004, Ch. 11 & 13) for the long run evolution of preferences in societies.

Experience

Experience with a choice situation brings an agent away from constructing her preferences and towards memory retrieval. She moves closer to the rational choice model of an agent simply revealing parts of a master list of preferences she has in her mind. Consequently, experience can result in choices being consistent.[15]

Hoeffler and Ariely (1999) argue that consumers construct their preferences in domains which are new to them, like for example soon to be parents when buying a baby stroller. Preferences eventually stabilize as the consumer gains experience in the domain. As a result, one can expect that the formation of prospect theoretic reference points is more prevalent in negotiations on a relatively unfamiliar topic than on an everyday commodity.

Decision Errors

A decision error—also called bias—is a deviation from a prescriptive decision-making model like utility maximization. If one allows for the possibility that preferences are constructed, that the process adapts over time, and that choices finally stabilize, then an assumed decision error might not be an error at all. What appears to be a mistake in the short-run can rather be part of a long-lasting trial-and-error improvement process. The agent might try out different heuristics to better judge their accuracy and effort and to better trade them off in the long-run.

Preference Uncertainty

Limitations of information processing and cognitive capabilities, as they are commonly assumed in models of bounded rationality, do not only affect information processing about consequences of actions. These limitations do likewise affect information processing about preferences for these consequences (March and Simon, 1958; Simon, 1973; March, 1978). This observation challenges the assumption inherent in many models of rational choice that preferences are precise and know with certainty. Theories of choice under ambiguity or conflict emphasize the

---

[15] A related point will be discussed in Section 3.2.1 with respect to whether market experience lessens the endowment effect.

complications of guessing (future) preferences (e.g. March, 1978, and Kreps, 1979).

A microeconomic model of *probabilistic preferences* is, for example, presented by Quandt (1956). Quandt assumes that a consumer is often ignorant of the exact state of her preferences, i.e. given two alternatives A and B, she is incapable of stating whether she prefers A to B, or B to A, or whether she is indifferent between the two.[16] According to Quandt, the average amount of ignorance concerning a particular alternative depends on the nature of the alternative itself and on the extent to which the agent can experiment with substitutes and thus acquire information, perhaps through conscious randomization of strategies. This ignorance is modeled via preferences over issues rather than alternatives and a probabilistic selection of issues that are taken into consideration when comparing two alternatives. As the selection of issues considered might (unconsciously) differ between choice situations, behavior becomes probabilistic and preferences are not generally transitive. Basing decision-making on a selection of issues rather than extensive information processing is in line with the psychological view of selective information processing in a heuristic decision-making. However, Quandt assumes utility maximization rather than heuristic decision-making.

From a more psychological perspective, (Fischer, Luce, and Jia, 2000; Fischer, Jia, and Luce, 2000) propose a model of *preference uncertainty* that deals with the feeling of ambivalence that arises from trading off different issues against one another in multi-issue decision-making. Like Quandt, Fisher et al. assume that (some) agents are not aware of their preferences when comparing (some) alternatives. In their model, preference uncertainty increases with within-alternative conflict: if a student chooses among different courses at university, for example, the choice is assumed to be straightforward if the most interesting course (issue 1) is the one with highest expected teaching quality (issue 2). In this case, there is no within-alternative conflict. If, on the contrary, the topic of a course A is clearly more interesting than the topic of another course B and the expected teaching quality of B is higher than of A, Fischer et al. assume that this within-alternative conflict makes the choice more ambivalent and the student tends to be uncertain about her preferences.

Fisher, Luce, and Jia (2000) define preference uncertainty as not being sure which of two alternatives one prefers, or to what degree.

---

[16] Note that ignorance does not imply that the agent is indifferent between the relevant alternatives but rather that she does not know her preferences.

Note that the former question is on the level of ordinal utility, whereas the latter question for the strength of preference is on the level of cardinal utility. Besides the within-alternative conflict of issues mentioned above, other reasons for this kind of preference uncertainty might be attribute extremity or factors not directly influenced by the choice set like familiarity with the domain, good, etc. Fischer et al. expect that preference uncertainty is higher if the choice situation is novel for the agent (cf. Hoeffler and Ariely (1999) and the baby stroller example given above); this implies that the uncertainty should decrease as preferences stabilize.

Preference uncertainty is—like preferences in general—not directly observable. It does, however, manifest in an agent's behavior: If an agent is for instance unsure about whether the positive features of an alternative outweigh its negative features, she is likely to firstly take longer time for evaluating it and to secondly express less consistent evaluations over time (Fisher, Luce, and Jia, 2000).

2.3.2 Empirical Evidence

Numerous studies put forward experimental evidence for constructive choice processes. Here, the discussion is limited to two examples: anchoring and response mode effects. See Bettman, Luce, and Payne (1998, pp. 199–207) for a review of other studies and Fischer, Luce, and Jia (2000) for experimental evidence concerning preference uncertainty.

Arbitrary Anchoring

Ariely, Loewenstein, and Prelec (2003) confronted subjects with six products without mentioning the market price. The products were computer accessories, wine bottles, luxury chocolates, and books. The study aimed at examining whether subjects have a predefined value for these consumer goods, or whether the valuation is constructed on the fly. To test for construction, the authors tried to influence the construction process by arbitrary anchors. If these anchors have an effect, then it cannot be the case that subjects have a predefined valuation that is simply retrieved from memory.

Subjects' willingness to pay for the goods was elicited in two ways: Firstly, they were asked whether they would buy each good for the last two digits of their U.S. social security number (which is essentially random) and, secondly, they had to state their willingness to pay (in dollars) for each product. A random device determined which good a subject might buy in the end and whether the social security number or the dollar amount defined the price.

For the analysis, subjects are grouped in quintiles with regard to their social security number. The quintile with the highest ending digits was, on average, willing to pay three times as much for a good, as the quintile with the lowest ending digits. For example, top quintile subjects stated to be willing to pay $56 on average for a cordless computer keyboard, whereas subjects in the bottom quintile expressed an average willingness to pay of $16 for the same good. The social security number anchors the subjects' mental process to either a high or a low number. The adjustment away from this anchor towards the willingness to pay is insufficient and, consequently, the numbers are different.

Besides this arbitrariness in absolute valuations, relative valuations were quite stable across subjects with different anchors. For example, 95% of the subjects valued a cordless keyboard more than a track-ball and so on. Ariely, Loewenstein, and Prelec (2003) reason that there might be a fundamental set of preferences which can be mapped on valuations. This mapping process can be manipulated by anchoring. Subjects did not really know how much they valued the items; at best, they had a range of acceptable values.

The authors further demonstrate, that anchoring effects are not restricted to money valuations of consumer goods. They provide evidence for the effect being observable for non-goods (specifically pain) and for durations (of pain) instead of money valuations. An interesting finding is that the anchoring effect does not diminish as subjects gain more experience with the goods, at least not with the degree of experience possible in this specific experiment.

Overall the sketched experiment shows that an arbitrary anchor influences subjects' willingness to pay for well-defined consumer goods. This confutes the assumption that valuations are simply retrieved from memory and favors the notion of a context-specific preference construction process.

Application to Negotiations

Anchoring is a well-known bias in negotiations. Studies on anchors in negotiations thereby oftentimes regard the initial positions of negotiators as anchors (Kristensen and Gärling, 1997b, 2000; Bazerman and Neale, 1992, Ch. 4).

Imagine, for example, you want to have your portrait taken by a street artist at a touristy place. You address one of the artists and ask for the price of a portrait. He pulls out a price-list and answers: '60 euros. But for you, my friend, I make a special price: 30 euros!'. You don't know this artist and he is not your friend. Why should he make a special offer for you?

The artist's intention is to anchor you at a high price of € 60. Even if you are perfectly aware that this price is outrageous for the quality offered, your adjustment away from this anchor is likely to be insufficient. Consequently, the price you pay—let's say € 15—will be higher than without anchoring.

Response Mode Effects (revisited)

A basic principle of rational choice models is procedure invariance—different modes in which a response is elicited should not influence the revealed preferences as long as the modes are strategically equivalent (cf. Sec. 2.1.2). One possible response mode is choice, i.e. deciding which of two alternatives is more desirable. Another response mode is matching, i.e. creating an alternative that is equally preferable as a given one. Even other response modes are bidding for alternatives or rating them. Preference reversals among these modes have been repeatedly demonstrated (Lichtenstein and Slovic, 1971; Lindman, 1971; Tversky, Sattath, and Slovic, 1988; Slovic, Griffin, and Tversky, 2002).

Nowlis and Simonson (1997), for example, tested preference reversals in a series of experiments. Their subjects were faced with purchase decisions for televisions, batteries, sunscreen lotion, hotel rooms, etc. All of the tasks had in common that there was one alternative with a high price and a prestigious brand name and a second alternative supplied by a lower ranked brand for a lower price. Thus, a purchasing decision requires trading off the price against the brand name and the quality implied by the brand name. An example is as follows:

- Color television A
  Brand: Sony; Price: $309
- Color television B
  Brand: Magnavox; Price: $209

In choice mode, both alternatives were presented and subjects were asked to choose which of the two they would buy. In rating mode subjects saw just one alternative at a time and were sequentially asked to indicate their likelihood of buying it on a scale from zero to twenty. From the two likelihoods one can construct a subject's implied choice. With procedure invariance due to simple memory retrieval, one should expect that subjects' responses are the same in both response modes.

The data suggests otherwise. Nowlis and Simonson (1997) present 51 tests involving different product categories, numbers of issues, etc. In 50 out of these 51 they find preference reversals—subjects tend to prefer the low-cost low-prestige alternative in the choice task and, to the contrary, they tend to prefer the high-cost high-prestige alternative in rating mode.

The explanation offered by the authors is as follows: some issues, like the numerical price, are easy to compare. They give an unambiguous ranking across alternatives. Other issues, like the nominal brand name, are difficult to compare precisely; the difference tends to be ambiguous and non-quantifiable. In choice mode, the weight an issue has is increased with its comparability—thus the low-cost television B is chosen. In rating mode, to the contrary, issues that are difficult to compare but have a rich context and association loom more important—thus the high-prestige television A is chosen.

In choosing, agents can apply issue-wise processing of alternatives. In rating, on the other hand, alternative-wise processing is required. The data of Nowlis and Simonson and other authors suggest that the trade-off across issues is influenced by these processing orders. Hence, procedure invariance does not hold and the evidence suggests that decision-making is a process including context and task dependent construction of preferences.

Application to Negotiations

In most negotiations agents evaluate offers received from their counterparty and, if an offer is not acceptable right away, they propose a counteroffer. These two activities are different response modes. Evaluating an offer is a binary choice situation: the offer is either accepted, or not. This choice likely includes comparing the offer to previous offers via issue-wise processing. Creating a counteroffer requires assigning values to issues via alternative-based processing. Hence, the relative importance of issues might differ between evaluating offers and creating offers.

Preference Stabilization

Coupey, Irwin, and Payne (1998) present data corroborating the notion of preference stabilization. Consumers tend to make less choice-matching preference reversals once they become familiar to applying a specific response mode to a product category. Furthermore, Cox and Grether (1996) study choice and valuation response modes. At the beginning of their experiment, they find strong evidence for preference reversals between choice and bidding in a second price auction. In later rounds, however, these preference reversals disappear.

The two studies support the idea that novelty and complexity in the domain, the elicitation procedure, or both are likely to result in preference construction. With experience and learning, decision-making becomes memory retrieval or it at least follows deterministic algorithms. Hence, preferences stabilize.

Summary

Both examples—anchoring and response mode effects—demonstrate that making a choice is more than retrieving preferences from memory, at least in some domains and for some agents. Decision-making is rather a task and context specific process of information acquisition, preference construction, and, finally, preference revelation. This process becomes more stable when the agent gains experience with the domain and the task.

2.4 Neural Basis of Preferences

The economic and psychological approaches presented so far try to understand and predict the working of preferences based on observed behavior. Cognitive neuroscience and neuroeconomics go one step further in unraveling the processes involved in decision-making and look inside the 'black box' human brain. These disciplines analyze human behavior and relate the activation of different regions of the brain to the economic tasks and choice situations (McCabe, 2003; Camerer, Loewenstein, and Prelec, 2005). Neuroeconomics is the least abstract approach to human decision-making discussed here.

There are several ways to identify specialized regions in the brain and relate them to different tasks: Early studies focused on descriptions of clinical disorders and experiments with patients that have brain lesions. Later on, neuroscientists developed invasive techniques for monitoring brain activity like inserting an electrode in the brain of an animal[17] and non-invasive techniques like imaging of brain activity.

The basic setup of an imaging experiment is to record brain activity of subjects performing either a treatment task or a control task. The difference of the images across groups indicates which brain regions are activated by the treatment task. Two imaging techniques frequently

---

[17] Note that the ethics of such invasive and potentially deadly techniques in animal experiments are highly debated.

used with human subjects in neuroeconomics are *Positron Emission Tomography* (PET) and, more recently, *functional Magnetic Resonance Imaging* (fMRI). Both techniques measure blood flow as a proxy for brain activation. For PET, radioactive isotopes are injected into the subject's blood and the emissions are imaged. Major disadvantages of PET are the radioactive substances and the relatively low temporal resolution. fMRI builds on the magnetic characteristics of blood and the changes due to oxygenation to image blood flow (Logothetis et al., 2001). With fMRI—currently the most powerful and most expensive brain imaging technique—one can record up to four images per second. Temporal and spacial resolution of these techniques are reciprocal.

Other imaging techniques that are of minor importance for the following overview are electro encephalograms (EEG), magneto encephalograms, and near infrared spectroscopy (NIRS). See Carter (1998) for an introduction to neurology in general and in particular for an overview on brain imaging techniques as well as the location of different brain regions referred to below.

The Biological Approach to Evolutionary Development

The evolution of the human brain began in ancient times when fishes formed a central system that pooled the endpoints of nerves from distant parts of their body. The nerves started pooling in different specialized modules. Some of these modules, for example, started being light sensitive and developed towards eyes and others connected to control movements. The result of this mechanic and unconscious reptile brain is still part of the human brain (brain stem and the cerebellum; Carter, 1998).

Building on the reptile brain, evolution brought up the more complex mammal brain also known as limbic system: the thalamus relates senses like seeing and hearing, the amygdala memorizes fear, the hippocampus performs this and other simple memory tasks, and the hypothalamus controls basic body functions. The mammal brain is, like the reptile brain, unconscious.

In the human brain the modules of the reptile brain and the mammal brain are still active—the limbic system creates, for example, emotions even if it is unaware of them. The by far biggest parts of the human brain are the cortex and the neocortex. They are divided into a left and a right hemisphere. Different regions and hemispheres of the cortex specialize on performing different tasks and thereby influence reasoning processes, behavior, and the perception of emotions.

Specialization

Neuroscientists were able to identify brain regions specialized to self-control, processing visual information or acoustic information, understanding languages, recognizing faces, controlling movements, performing spatial or abstract reasoning, mathematical computations, future planning, remembering the past, perceiving positive emotions like happiness and negative emotions like fear, teariness, and disgust, recognizing spiritual or religious phenomena, etc.

Microeconomic models assume that people have a fixed set of preferences and aim at satisfying them. The abstraction of this unified account of behavior is a master list of preferences in the mind that simply has to be retrieved from memory. If this corresponds to the organization of the brain, one would expect that in a choice situation one specific brain region that specialized on storing preferences is activated.

Behavioral models that account for the construction of preferences via different heuristics and algorithms, on the other hand, would predict different brain regions to be active during the construction of preferences as decision-making might be performed in different brain regions depending on the specific context and task.

Neuroeconomic Experimental Evidence

Using PET, Parsons and Osherson (2001) found, for example, that inductive reasoning and guessing probabilities mostly activates the left hemisphere while deductive reasoning for answering logic questions is mostly performed in the right hemisphere. In another experiment, Fiorillo, Tobler, and Schultz (2003) demonstrate that the level of dopamine—a chemical that functions as neurotransmitter—in the brain increases during gambling. Furthermore, the dopamine level depends on the probability and the magnitude of a potential reward. This leads the authors to the hypothesis that the dopamine level itself is a rewarding property of gambling. This would explain people gambling in, for example, casinos although they know that the odds are against—they are not solely motivated financially or by misperceiving probabilities, but they get physical pleasure from uncertainty and risks.

Smith et al. (2002) relate behavioral and neural effects of evaluating gains and losses. In their PET imaging experiment, subjects had to chose among lotteries in a 2×2 design for the belief structure—either risk, i.e. known probabilities, or ambiguity, i.e. without knowledge of the probabilities—and the payoff structure—either gains or losses. The behavioral data resembles frequent empirical findings: in line with prospect theory's s-shaped value function subjects tended to be risk-taking for losses and risk-averse for gains when probabilities were known. They were ambiguity-averse for gains and losses.

The brain regions activated during decision-making resemble this behavioral interaction effects. The authors find evidence for two separate but interacting choice systems with sensitivity to losses: one for processing losses from risky lotteries and one for the other three stimuli. The identification of the 'risky-loss system' as a neocortical dorsomedial system leads them to the conjecture that this system uses more calculation. The processing of the other three stimuli, on the contrary, is done in a ventromedial system, a part of the evolutionary older mammal brain—here, decisions tend to base on visceral representations and instincts.

A similar point is made by McClure et al. (2004) in an experiment using fMRI: their subjects use different brain regions to evaluate short term and long term monetary rewards. More specifically, all intertemporal choices exhibit about the same activation of the lateral prefrontal and parietal areas, i.e. brain regions that are usually associated with deliberate reasoning including numerical computations. This appears to be in line with normative economic models for intertemporal choice and discounting of future payoffs. But for near-term rewards, there is an activation of additional brain regions; immediate rewards tend to be additionally evaluated in parts of the limbic system, i.e. a more affective and evolutionary older part of the brain that is consistently associated with impulsive behavior. This speciality of immediate rewards with respect to its neural processing can be used to explain what appears as anomalies from a microeconomic discounted utility perspective; see e.g. O'Donoghue and Rabin (1999) and Frederick, Loewenstein, and O'Donoghue (2002) for reviews of the economic theory and empirical evidence on intertemporal choice.

The aforementioned studies demonstrate that neuroeconomics can relate behavior in some standard economic experiments to the internal functioning of the brain. The results suggest that decision-making is not simply retrieving preferences from a specific part of memory but that it is a process with task-dependent activation of different brain regions.

Summary

The purpose of this section is not to lay the foundations for analyzing brain functions in a negotiation. It rather is to briefly review some insights neuroeconomics provides on preferences and individual decision-making. Overall, several neuroscientific studies show that human behavior is oftentimes determined by a competition between lower level, automatic processes that reflect evolutionary adaptations to specific environments, and the more recently evolved, uniquely human capacity for abstract, domain-independent reasoning and planning (McClure et al., 2004). Thus, decision-making is a complex interplay of different brain regions. The psychological view of decision-making as a task- and context-dependent process that sometimes involves non-optimizing impulsive behavior seems to have a higher congruence with reality than the microeconomic view of unboundedly rational utility maximization.

2.5 De Gustibus non est Disputandum

The previous section presented a model of constructive preferences and empirical evidence corroborating this notion. The entire idea that preferences might be labil instead of being a stable and persistent characteristic of any agent is in contradiction to the rational choice model outlined in Section 2.1 and is highly disputed by many economists. Their assumption is that preferences do not change; either production technology or information changes, but preferences are stable. Obviously, most (if not all) economists grant that this assumption is not in congruence with reality; they do, however, insist on it to retain analytical tractability and coherence of economic theory. This section aims at bringing together the two perspectives and evaluating which one is more appropriate for the subsequent study of negotiators' behavior.

2.5.1 Stability of Preferences

One of the most forceful argumentations for assuming stable preferences has been made by Stigler and Becker (1977). They title their paper 'De gustibus non est disputandum' and offer two interpretations of the proverb: Firstly, it can stand as advice to end a dispute once it has been resolved into a difference of taste. The rational for this interpretation is that tastes are unchallengeable axioms of human behavior. Everyone possesses individual tastes, i.e. preferences. They may change endogenously but cannot be altered exogenously, for example, by persuasion. The second interpretation is preferred by Stigler and Becker: tastes neither change capriciously over time nor do they differ distinctively between people. They are stable facts for everyone.

The first interpretation takes a behavioral viewpoint. With this, an economic model can explain behavior up to preferences, but not further. Once preferences enter the picture, the analysis has to be handed over to other disciplines like psychology and anthropology. The second interpretation, on the other hand, is a traditional economic perspective. It allows economists to fully explain behavior and differences among people without having to hand the analysis over to other behavioral sciences.

One can immunize the notion of stable preferences by simply incorporating the entire history and context as issues in a decision problem. This is exactly the way Stigler and Becker (1977) take. The utility function that is maximized by consumers in their model does not only have the amount of consumed goods as parameter, but as well the *production technology* available to the respective agent and an individual factor termed *human capital*.

Human Capital

The human capital entering the utility function is an accumulation of everything the agent has consumed and done over her entire life. In examples given by Stigler and Becker, it is always restricted to the consumption in one domain but in the general model it is the agent's entire life story. If an agent becomes addicted to, for example, drugs or good music, then her preferences for these goods do not change. It is not that this year she likes good music more than she did five years ago. Instead, she gets a higher utility from listening to music now than she got from it five years ago because her utility is defined over listening to it given the fact that she attended six operas last year, five concerts the year before, etc.

Modeling human capital as input to the utility function is an axiomatic approach: It allows to explain any behavior. It is extremely unlikely that two agents with equivalent human capital face the same choice situation (and if they do so, it is even more unlikely that an economist is around to observer choices). Furthermore, one and the same agent can never face the same situation twice, as decisions are dated and her human capital changes. Thus, any behavior can be explained by the general model as long as one does not impose a special functional form on the utility function. The drawback of this powerful modeling device is that tractability of modeling deteriorates as the utility function gets complex and the information requirement is enormous.

Production Technology

A challenge to the notion of stable preferences is the question how agents can have preferences over goods and technologies that do not yet exist. Today, cars are an ubiquitous device used in transportation. People oftentimes seem to prefer the car to many other modes of transportation. But how about the time before their invention in the nineteenth century?[18] Did people have a preference for using a car even if they did not know it would be invented?

The modeling device used to overcome this is production technology. People do not directly have a preference for using a car—they have a preference for transportation. Transportation via horse carriage is fine if this is the most suitable production technology at hand. If however, a new technology is invented, it might subsidize the usage of an older technology in some situations. The underlying preference for transportation is satisfied anyways.

Incorporating production technology in the utility function is a potent way to explain long-term changes in behavior without adhering to notions of changing preferences or preference evolution. Explaining short-term phenomena like framing, anchoring, and response mode effects, on the other hand, is much more difficult.

Information

A third factor economists oftentimes use to explain what might be accounted to preference changes is the information available to agents. Information is omnipresent in economic models as agents base their beliefs and hence decisions on information regarding the available actions, the potential outcomes, probabilities, other agents' behavior, etc. Thus, many behavioral patterns can be explained by making assumptions on the agents' information and beliefs.

In a recent article, Ariely, Loewenstein, and Prelec (2006) report on experimental data and conclude that many agents do not have a fixed monetary value for a good and that in some cases agents do not even know whether they like or dislike an experience. The authors' conclusion is that there is no such thing as a fundamental value or fundamental preferences.

Ariely et al. conducted a classroom experiment and told their students that their professor, i.e. Dan Ariely, would like to give a 15-minute poetry reading. Half of the students were asked whether they would be willing to pay $ 2, the other half was asked whether they would attend the reading if they received $ 2. Afterwards, all subjects were told that the recitation would be for free and they could sign up if they wanted

---

[18] One could argue that there were car-related transportation devices since about 100 before Christ. However, the assumed change in preferences can easily be stretched to this longer time span.

to be notified via e-mail. If asked to pay, only 3% wanted to attend the poetry reading. On the other hand, 59% were willing to attend it if they would be paid. Furthermore, 35% who thought they would have to pay wanted to attend for free whereas just 8% liked the idea of a free poetry reading after being offered a payment for it in the first place. The difference is highly significant. The authors conclude that the initial question influences subjects' responses and that there is no fundamental value.

A counter-argument is that this results can easily be explained by information and beliefs. The experiment can be seen as a signaling game. Students do not know whether the reading will be good or bad but they assume that their professor knows it. Under the assumption that the professor wants a full classroom as audience, his willingness to pay signals a bad quality whereas asking the students to pay signals a good quality. The signal is used by students to update their beliefs about the quality and based on this they make their decision. Thus, different attendance rates are not surprising and not necessarily due to a change in preferences caused by the initial question.

To rule this explanation out, Ariely et al. conducted another closely related experiment. Students were given a one minute free trial to personally judge the quality of their professor's poetry readings and the assignment to different treatments was made public. Each student received the same sheet of paper with two questions. Assignment to one of the two questions was essentially random via the last digit of the U.S. social security number and each student knew that some others faced the other question. The results are similar to the ones cited above.

It does not appear plausible that subjects can infer information from public random assignment to treatments. Nevertheless, across treatments there are differences in the evaluation of the poetry reading. These differences cannot be due to production technology or information. They could be explained by accumulated human capital as it might be that the mental effort devoted to understanding the different questions is different across groups. However, this appears rather contrived compared to the behavioral explanation that deciding on the attendance of the reading involves a heuristic preference construction process that is influenced by anchoring.

Changing Preferences

Economists tend to dislike changing preferences as this could explain each and every decision—predicting behavior becomes impossible. Remember that rational choice is defined as a pattern of choices that can be explained by an underlying preference relation. If this preference relation can change after each choice, there can be no irrationality and thus no sharp-edged notion of rationality that would give some bite to a model of decision-making. Economists like Stigler and Becker (1977) argue against the behavioral perspective, as changing preferences allow for endless degrees of freedom in behavior.

Supporters of preference construction and adaptive decision-making theories argue that their models are descriptively more valid and are closer to how agents really make decisions. Research on preference construction has made enormous progress over the last decades and the purpose is not to allow for arbitrary preference changes—this would indeed erase any predictability as argued by Stigler and Becker. The idea rather is to find—if possible—well defined rules, patterns, and heuristics how people behave and to apply them in the domains where empirical evidence suggests that they are appropriate. The same behavioral principles, like loss aversion for example, have been applied in a variety of economic situations: stock markets, labor economics, consumer goods, horse race betting, purchases of insurance, etc. (see Camerer, 2001, for an overview). However, this is not (yet) as broadly applicable as rational choice theory, since it is not a compact and coherent theory 'system'.

Preferences in Prospect Theory

In prospect theory, outcomes are judged in relation to a reference point. This reference point can change over time, even if prospect theory itself does not explain the nature of the change. In the view of Stigler and Becker (1977) one could incorporate this reference point in the utility function and the location of the reference point in the human capital which in turn is part of the utility function.

The first step, i.e. having a utility function that depends on a reference point, is exactly what prospect theory does. It thereby retains some congruence with conventional economic utility maximization models (Kahneman and Tversky, 1979; Tversky and Kahneman, 1991, 1992).

Adding the origin of the reference point to the utility function as well, however, does not help if one does not simultaneously specify how exactly each possible human capital determines the reference point. Here it appears more tractable to base models on heuristics dealing with the perception of choice tasks and the formation and adaptation of reference points (Loewenstein and Issacharoff, 1994; Strahilevitz and Loewenstein, 1998).

2.5.2 Summary

'Economic theory, since it has been systematic, has been based on some notion of rationality.' (Arrow, 1986, p. 388). According to Arrow (1986), the currently prevailing notion of rationality in economics is utility maximization. However, it is not necessarily the only viable view on individual decision-making. Roth (1996) categorizes models on human decision-making in five broad categories. Four of these five correspond to the different approaches sketched in Figure 2.1 and were presented in more detail throughout the chapter.

Payoff maximizing economic man: According to this approach, agents choose the highest expected monetary payoff they can get. The model is still used as approximation in contemporary economic research, although its shortcoming is widely acknowledged: The model can, for example, not explain why agents would buy insurances.

Utility maximizing economic man: The descriptive limitations of expected payoff maximization led to an extension of the model towards expected utility maximization. Bernoulli proposed it in the 18th century as solution to the St. Petersburg paradox and it allows explaining the usage of insurances. (Cf. Sec. 2.1)

Almost rational economic man: This is the name Roth uses for non-expected utility theories like prospect theory that retain standard microeconomic assumptions like preferences and utility optimization and 'just' extends expected utility theory to account for systematic biases. (Cf. Sec. 2.2)

Psychological man: According to this psychological approach, an agent does not have a list of preferences where she can look up what she prefers, but has a set of mental processes and heuristics that are task and context specific. (Cf. Sec. 2.3)

Neurobiological man: Recent neuro-scientific research suggests that humans do not (even) have a fixed collection of mental processes and heuristics like the psychological approach suggests. Instead, decision-making is guided by biological and chemical processes. (Cf. Sec. 2.4)

These categories are used by Roth (1996) to answer the question why theoretical economists are reluctant to depart from the rational model, despite considerable contradictory evidence? He argues that mainstream economics changed from analyzing payoff maximizing man to utility maximizing man not only because of isolated phenomena like the St. Petersburg paradox, but because there were very large phenomena like entire industries (insurance) and large markets (future markets). The importance of these systematic deviations from the simpler model led to utility maximization as more accurate approximation, i.e. to a theory with higher congruence to the real world. The extension certainly was revolutionary but it allowed maintaining coherence of economic theory, as payoff maximization is a special case of risk-neutral utility maximization and utility maximization is analytically tractable.

Many experimental economists, behavioral economists, and cognitive psychologists suggest extending the standard modeling approach even further to get even higher congruence with reality. However, the neuro-scientific approach shows that the models proposed are all 'just' simplifying approximations. Furthermore, behavioral models are more complex than traditional economic models and involve more unobservable factors—thus, coherence deteriorates. There is no correct, right, or best way for trading off a theory's abstract properties congruence and coherence. More specifically, a theory's characteristics like generality, manageability, tractability, realism, and predictive accuracy would have to be traded off for deciding which approach is best—as this is impossible on a general basis, different approaches to human decision-making co-exist. Behavioral economics is an attempt to integrate the coherence of traditional economics and the congruence of psychology.

Outlook

An unanimous decision whether microeconomic or behavioral theories are better suited for modeling, analyzing, and predicting behavior negotiations cannot be reached. A deviation from the strict microeconomic notion of exogenously given and invariable preferences might be useful *if* there are behavioral models that predict the precise nature of preference changes and *if* these models are supported empirically in negotiations.

Chapter 3 points out how a negotiation process might influence negotiators' preferences. The corresponding empirical tests are reported in Chapters 4 and 5.

3

Preferences in Negotiations

> *When two or more parties need to reach a joint deci-*
> *sion but have different preferences, they negotiate. They*
> *may not be sitting around a bargaining table; they may*
> *not be making explicit offers and counteroffers; they may*
> *even be making statements suggesting that they are on the*
> *same side. But as long as their preferences concerning the*
> *joint decision are not identical, they have to negotiate to*
> *reach a mutually agreeable outcome.*
>
> (Bazerman, 2006, p. 133)

As Bazerman points out, in a negotiation two or more parties are interested in reaching one of several possible agreements, but their preferences over these agreements are not completely identical. In multi-issue negotiations, studied here, parties usually have the possibility to simultaneously negotiate over several issues and to search for integrative potential. Negotiators play a non-constant-sum game.

This Chapter first briefly reviews research on negotiations in Section 3.1 to pinpoint different approaches to assessing preferences and behavior in negotiations as well as common biases found in negotiation behavior. Subsequently, Section 3.2 discusses the assignment of property rights and expectations as two potential causes for reference points and shifts of reference points. Section 3.3, finally, introduces the *attachment effect* in negotiations, i.e. a systematic effect of offers in a negotiation on the parties' preferences. This effect builds on the role of expectations for the formation of reference points and relates to the negotiation analytic research surveyed in Section 3.1.

3.1 Negotiations

Negotiations are non-individual decision-making processes. More specifically, they can be defined as follows: Negotiation is a decision-making process involving two or more parties that jointly resolve a dispute or determine outcomes of mutual interest via exchanging ideas, arguments, and offers. Parties thereby can be individuals, groups, organizations, or computer-based decision-making models. The dispute arises from the fact that no party can achieve its objectives without the agreement of someone else and outcomes involve resource allocations as well as courses of action to take in the future (Gimpel et al., 2003).

The preceding perspective on negotiations is a rather broad characterization of a communication process. In a more precise definition, the following features characterize negotiations: (1) agents believe that they have conflicting interests; (2) intermediate solutions or compromises are possible; (3) communication is possible; (4) parties may make provisional offers and counteroffers; and (5) offers do not determine outcomes until they are accepted by both parties (Kristensen and Gärling, 1997a).

In the present study, the focus is on a specific set of such negotiations additionally characterized by the following features:

1. Negotiations are *bilateral*, i.e. they involve two parties that have to reach a compromise agreement. Furthermore, for most of the following discussion, parties are assumed to be monolithic—each party is a single individual and not, for example, a company or a nation.
2. The two parties aim at finding an agreement on *multiple issues*, i.e. there are one or more objectives over which the parties negotiate.
3. Negotiators have *incomplete information* on their counterparty's preferences over different possible agreements. They might, for example, know the direction of monotonicity of preferences but not all trade-offs between different issues at all levels.

3.1.1 Interdisciplinary Research

Negotiations are studied in several disciplines: economics, decision science, psychology, information systems, computer science, management science, political science, law, anthropology, sociology, etc. Bichler, Kersten, and Strecker (2003), for example, provide an overview on the interdisciplinarity of negotiation research and contributions coming from the different fields. In the following, only concepts and ideas from economics, decision science/ negotiation analysis, and information systems are taken into account as these disciplines (1) provide the most formal approaches to negotiations, (2) are most closely related to the theories on preferences outlined in Chapter 2, and (3) are most directly related to bilateral, commercial, multi-issue negotiations.

Game Theory

Game theory is an economic framework for thinking about strategic interaction; it directly builds upon rational choice theory. As a negotiation is a strategic interaction, game theory is commonly used by economic theorists to analyze negotiations or *bargaining games* as they are frequently termed in this context. The following overview on game theoretic analysis of bargaining models does by no means strive for completeness. It rather is a very short introduction to the different approaches game theory uses for assessing negotiations. See Roth (1985) for a collection of articles on game-theoretic bargaining models in general, Thomson (1994) for an overview on cooperative bargaining models, Binmore, Osborne, and Rubinstein (1989) for an introduction to non-cooperative bargaining models, and Ausubel, Cramton, and Deneckere (2002) for a review of studies on incomplete information in non-cooperative bargaining models.[1]

The abstract bargaining problem mainly considered in game theory is that two 'individuals have before them several possible contractual agreements. Both have interests in reaching agreement but their interests are not entirely identical. What "will be" the agreed contract, assuming that both parties behave rationally?' (Rubinstein, 1982, p. 97). Different models differ in the rules of the interaction, i.e. the agents' possible strategies, for resolving this bargaining problem.

Cooperative Game Theory

The first game-theoretic approach to bargaining games was cooperative game theory: It analyzes games in which the players can engage in coalitions and arrange side payments to come to a mutual and enforceable agreement. Agents are assumed to act cooperatively to come to a beneficial outcome. They might, for example, have to split a fixed amount of money and have conflicting interests in the share they receive. Nevertheless, both have a common interest in dividing the money at all so that they, or at least one of them, receive any money.

---

[1] Furthermore, see Fudenberg and Tirole (1991) for an extensive introduction to game theory in general.

Axiomatic approaches to single-issue negotiations are not concerned with the specific offers agents make during a negotiation but rather with the question which share an agent might reasonably claim for herself and which outcome is fair or just. Nash (1950) proposed one of the first axiomatic bargaining solutions. It bases on the axioms (1) symmetry, (2) Pareto optimality, (3) invariance to equivalent utility representation, and (4) independence of irrelevant alternatives. It turns out that these four axioms uniquely determine a solution which Nash characterizes as fair and which is assumed to be acceptable by rational agents. Besides the Nash solution, there are numerous other solution concepts in cooperative game theory, e.g. the Kalai and Smorodinsky (1975) solution or the Gupta and Livne (1988) solution.

Cooperative approaches do not offer a description how agents are supposed to reach the agreement given by the solution. Nor do they detail how the agents' cardinal utility functions that are necessary for calculation can be assessed. The axiomatic solutions rather constitute agreements that are supposedly fair with respect to different notions of fairness and that should be acceptable for rational agents given their utility functions. They are suggestions an omniscient mediator could put forward or an arbitrator could impose.

The analysis of bargaining situations was dominated by such axiomatic approaches from the 1950s on. However, Nash (1951, 1953) already proposed to abandon coalition formation, communication outside the game, transferability of payoffs via side payments, and enforceable agreements. The so called *Nash program* is a research agenda aiming at basing axiomatic solutions of cooperative games on non-cooperative equilibria. The non-cooperative, strategic analysis of negotiations finally matured at the beginning of the 1980s (e.g. Rubinstein, 1982; Fudenberg and Tirole, 1983; Fudenberg, Levine, and Tirole, 1985). With this, the necessity for an omniscient arbitrator is annulled and the process of negotiating is formally captured in game theoretic models.

Non-Cooperative Game Theory

Cooperative models deal with the outcome of a negotiation. On the contrary, non-cooperative models are concerned with the process of how to reach the outcome. Both are different ways to solve the same game; they are complementary approaches corroborating each other. In a seminal article advancing the non-cooperative analysis of bargaining games, Rubinstein (1982) presented a model of a multi-period, infinite-horizon bilateral bargaining game with alternating offers on the division of a good. The value of the good diminishes from period to period and both players have perfect information on the rules of the game, their preferences, and their offers. Under these assumptions, Rubinstein shows that there is a unique subgame perfect equilibrium—one of the main structural characteristics is that patience gives bargaining power and a striking result is that the two parties will reach an agreement with the first offer. Thus, in equilibrium there is no alternating offer exchange as this would delay an agreement and therefore diminish the value to be distributed between parties.

The analysis of bargaining games becomes way more difficult under incomplete information, i.e. if either one or both bargainers do not know their counterparty's time preferences or valuations for a good, for example. In this setting, the equilibrium depends on the beliefs of bargainers and equilibria oftentimes do not guarantee efficiency. Early analyzes of one-shot sealed-bid bargaining situations (also called *static bargaining*) under incomplete information are presented by Myerson and Satterthwaite (1983) and Chatterjee and Samuelson (1983); models of alternating offer bargaining (also called *sequential bargaining*) under incomplete information are given by, e.g., Rubinstein (1985) and Fudenberg, Levine, and Tirole (1985). If just one of two bargainers has private information and the other one has all the bargaining power, one can come to unique equilibrium models. However, if either both bargainers have private information, or if they alternate in proposing offers, then a multiplicity of equilibria arises (Fudenberg and Tirole, 1991, Ch. 10). See Ausubel, Cramton, and Deneckere (2002) for a wide overview on bargaining under incomplete information.

Especially challenging is the case of alternating offer bargaining with two-sided incomplete information: the problem is that the negotiators can use their offers and cause delays in bargaining to (untruthfully) signal their preferences and influence their counterparty's beliefs. A player might, e.g., aim at signaling that she is a buyer with a very low valuation (thus demanding a low price) or a very patient player (thus demanding a large share of the pie as delay is not too costly for her). This signaling results in inefficient delays and, depending on the specific model, in a no trade theorem, i.e. the probability of trade might converge to zero (Ausubel and Deneckere, 1992). See Ausubel, Cramton, and Deneckere (2002, pp. 1934–1936) for a review of the sparse literature on sequential bargaining with two-sided incomplete information.

Multi-Issue Negotiations

Under complete information, i.e. if both negotiators' preferences regarding all possible outcomes and their strategy spaces are common knowledge, a multi-issue negotiation reduces to a single-issue negotiation. Knowing preferences implies that the set of all Pareto optimal agreements (the so called *Pareto frontier* or *contract curve*) can be calculated. Bargaining then 'only' is agreeing on a single point on this Pareto frontier which is the same as in a single-issue negotiation where each agreement is Pareto optimal.[2] Thus, under complete information cooperative as well as non-cooperative bargaining models can be applied to multi-issue settings. Complete information might, for example, be assumed in a game where parties negotiate over the elements of a product bundle, competitive market prices for each of the products exist, and both parties are assumed to evaluate each product with its market price.

Under incomplete information, multi-issue negotiations do not reduce to single-issue negotiations. Besides the general difficulty that both parties try to (untruthfully) signal their preferences and influence their counterparty's beliefs, the additional difficulty that they do not know the Pareto frontier enters the game. Depending on the solution concept, one can find equilibria in alternating offer multi-issue negotiations under incomplete information: with an infinite agreement space there are, for example, infinitely many Nash equilibria. As an example, consider the following strategy with a focal agreement x^*: A negotiator rejects any offer except x^*, and offers x^* each time it is her turn to make an offer. Both parties in a bilateral negotiation playing this strategy is a Nash equilibrium if x^* gives either party higher utility than any outside option: No party has an incentive to unilaterally deviate from this equilibrium. Playing any other strategy would solely delay agreement (potentially infinitely long) without changing it. Thus, in the absence of time preferences (and even more so if future consumption is discounted), the deviating party does not gain an advantage by a different strategy.

The same problem of selecting among infinitely many Nash equilibria occurs in the relatively simple case of a single-issue alternating offer bargaining game under complete information. However, for this game Rubinstein (1982) was able to show the existence of a unique sub-

---

[2] An additional element that might enter strategic considerations (depending on the specification of the game) is that parties might use deviations from the Pareto frontier as threats. With risk aversion, this could have an influence on the single-issue negotiation.

game perfect equilibrium among the infinitely many Nash equilibria. To date, there is no such meaningful refinement of the Nash equilibrium concept for multi-issue alternating offer negotiations under incomplete information. Instead, these games are oftentimes analyzed by means of negotiation analysis (see Sec. 3.1.1 below).

Summary

Game theory builds on rational choice theory and studies strategic interactions like bilateral negotiations. Like rational choice, game theory mainly follows the coherence theory of truth. Game theoretic equilibrium models have pinpointed several features of bargaining situations and the influence of negotiators' characteristics like their time preferences. However, with respect to bilateral alternating offer multi-issue negotiations under incomplete information, the application of game theory has (to date) two major drawbacks: (1) there is no solution concept that suggests a single equilibrium, a small set of equilibria, or a set of reasonably homogenous equilibria and (2) even if one would exist, equilibrium models assume rationality of all agents and, thus, do not offer much in case one or both parties do not act rationally as numerous empirical studies suggest.[3]

The behavioral approach to negotiations that follows next is much less concerned with a coherent model of decision-making in strategic situations but focuses on congruence with real decisions. Based on this description, advice for negotiators can be deduced even in the absence of an equilibrium model or omnipresent rationality.

Negotiation Analysis

Prior to 1982, description of negotiation behavior mainly originated from psychology and normative models on behavior in negotiations were restricted to game theoretic studies. In an influential book Raiffa (1982) merged the two distinctive fields to a common *asymmetric prescriptive/descriptive approach*. He suggested to advice the focal negotiator conditional on the best estimate of her cognition and behavior as well as her counterparty's cognition and behavior that can be obtained from descriptive research. The field of research Raiffa started with this idea is termed *negotiation analysis*; it mainly involves game theory, social psychology, cognitive psychology, and (multi-criteria) decision theory (Raiffa, 2003).

---

[3] Noteworthy exceptions are equilibria in dominant strategies that do not depend on the counterparty's rationality. However, there certainly is no equilibrium in dominant strategies for the negotiations considered here.

Descriptive Basis

One approach to describing negotiator behavior is social psychology. Social psychology is the study of how thoughts and behavior of individuals are affected by the presence of other individuals—it studies how mental and social processes interact. According to Bazerman et al. (2000), research in social psychology can be classified in two subdomains: individual differences of negotiators and situational characteristics. Individual differences summarize demographic characteristics like age and gender and personality variables like positive or negative self-conception, conformity, and intelligence. See e.g. Pruitt and Carnevale (1993) for a review of this field of research. Situational characteristics, on the other hand, summarize the structural variables of a negotiation like incentives, bargaining power, deadlines, third parties, and the question whether parties are monolithic or not. Thus, situational characteristics are close to the factors studied in game theory; one difference however is, that game theory studies these factors in an abstract, theoretical way and social psychology studies them empirically.

A second approach that builds the descriptive basis of negotiation analysis is behavioral decision research. This field studies the systematic ways in which negotiators (or decision-makers in general) deviate from rationality or prescriptively optimal behavior. Behavioral decision research allows researchers to predict a-priori how agents will make decisions that are inconsistent, inefficient, and base on normatively irrelevant information (Bazerman et al., 2000).

Behavioral decision research assumes that decision-makers base their decisions on heuristics (cf. Sec. 2.3). Mostly, these heuristics produce 'correct' results with relatively low effort. Thus, in an accuracy-effort framework, it is rational to apply heuristics. In some situations, however, the heuristics that are oftentimes beneficial lead to unintended, systematic biases. The study of biases is not meant to understate the capabilities of human decision-making but biases are, on the one hand, the instances of human behavior where it is easiest to comprehend the underlying heuristics and, on the other hand, their occurrence most directly suggests advising the decision-maker how to overcome the bias.[4]

---

[4] A list of common biases in negotiations that reviews work in behavioral decision research is given in Section 3.1.3 below.

Prescriptive Advice

The advice negotiation analysis can offer to a focal negotiator is twofold: it concerns the negotiator herself and the belief about the counterparty. With respect to the negotiator herself, prescription mainly intends a debiasing. The purpose is to help a negotiator in realizing and overcoming a bias she is prone to. To this end, negotiation analysis emphasizes the importance to prepare for a negotiation. This process includes becoming aware of one's best alternative to a negotiated agreement abbreviated *BATNA* (What happens if the negotiation fails?) and assessing the trade-offs between issues in a multi-issue negotiation. A typical advice is: 'To be fully prepared to negotiate, you must clearly identify your priorities. Effective trade-offs can then be accomplished by conceding less important issues to gain on more important issues.' (Bazerman and Neale, 1992, pp. 70–71). While such simple rules appear quite obvious, numerous studies report that negotiators oftentimes act differently.

With respect to the beliefs about the counterparty, negotiation analysis on the one hand advises to gather information on the counterparty and, on the other hand, to be aware of the biases that might affect the counterparty. If the other side does not act rationally but is prone to framing effects, for example, reframing one and the same offer might decide whether it is acceptable or not.

Extensive lists of prescriptions for negotiating rationally are given by Raiffa (1982), Fisher and Ury (1983) Bazerman and Neale (1992, Parts II & III), Raiffa (2003), and Bazerman (2006, Ch. 9).

Summary

Negotiation analysis combines descriptive (mainly behavioral and psychological) research with prescriptive (mainly economic) advice how to negotiate rationally. This advice is usually directed to one of two negotiators, the so called focal negotiator. Several authors are, however, eager to point out that this advice does not only serve the focal negotiator but can increase both parties' gains from an agreement as negotiation analysis helps finding integrative potential.

Negotiation analysis is commonly applied to studying decision-making in bilateral multi-issue alternating offer negotiations under incomplete information where game-theoretic bargaining models lack the existence of meaningful equilibria.

3.1.2 Negotiation Process Model

At the beginning of Section 3.1, negotiations were characterized as non-individual decision-making processes. This process dimension is detailed in the following to better understand the single steps of negotiation decision-making and to identify the mental processes in which negotiators are prone to different biases. The process perspective is not absent in the research presented so far: models in non-cooperative game theory oftentimes view negotiations as extensive form games and thus as processes and negotiation analysis handles the preparation for a negotiation as separate step before starting the negotiation. However, handling a negotiation as communication process is studied most extensively in the information systems literature.

Negotiation as a Communication Process

The Media Reference Model by Schmid (1999) and Lechner and Schmid (1999) structures a transaction process along four phases of interaction: the information, intention, agreement, and settlement phase. This abstract transaction process can be applied to negotiations as a subclass of general transactions. In the Montreal Taxonomy, Ströbel and Weinhardt (2003) refined this structure specifically for electronic negotiations. The refinement contains sub-phases in the intention and agreement phase as sketched in Figure 3.1 (cf. Ströbel and Weinhardt, 2003, Fig. 1 & 2). The intention phase includes offer specification, submission, and analysis. The agreement phase comprises offer acceptance or rejection. In case of rejection, the process goes back to the intention phase. Messages exchanged in this negotiation process are advertisements, offers, and contracts.

Agreement and Negotiation Processes

The Montreal Taxonomy sees negotiation processes as special cases of agreement processes and points out that an agreement can be reached without a negotiation. This is the case if the above process is executed without going back from the agreement phase to the intention phase—a 'negotiation process takes place when the first agreement phase fails' (Ströbel and Weinhardt, 2003, p. 146).[5] According to this definition, a process can only be characterized as a negotiation ex-post as it is necessary to know whether the first offer was accepted or countered with

---

[5] According to this definition the dictator game would, e.g., not be classified as negotiation. Many game theorists do, however, see the dictator game as one—admittedly very rudimentary—bargaining game.

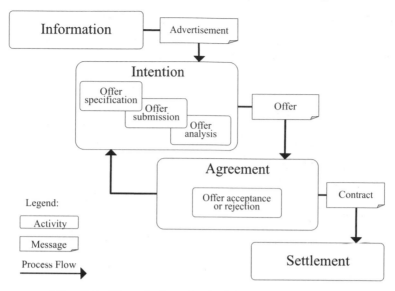

Fig. 3.1. Negotiations in the Montreal Taxonomy

another offer. However, ex-ante a negotiation only requires that the first agreement phase can in general be followed by a second intention phase. If this really happens in a specific communication of two agents is not the pivotal question.

Besides the phases presented in Figure 3.1, the Montreal Taxonomy proposes offer matching and offer allocation as further sub-phases within the agreement phase. These additional phases are specific to auctions and irrelevant in (other) bilateral negotiations. Thus, they are omitted here.[6]

Refined Process Model

Besides the phase structure of the Montreal Taxonomy, Jertila and Schoop (2005) propose the differentiation between private information

---

[6] Auctions are widely used class of market mechanisms and the differentiation to negotiations is not always clear cut. To avoid confusion, it is briefly discussed here. Auctions are market mechanisms with an explicit set of rules determining resource allocation and prices based on bids from the market participants (McAfee and McMillan, 1987). Thus, auctions are a special subset of negotiations as they satisfy the above definition of a non-individual decision-making process. Arguments are rare in auctions, but agents resolve a dispute on the allocation of resources by communicating via offers. The difference from auctions to other negotiations lies in the specification of the protocol to be followed: Auctions are negotiations with a well specified and enforceable protocol. Furthermore, they are almost always non-bilateral processes and, thus, not further considered in the present study.

and public information in negotiations. Accordingly, the process of the media reference model and the Montreal Taxonomy can be refined: The detailed version is displayed in Figure 3.2. The labels on the left hand side give from top to bottom the four phases of interaction as introduced in the media reference model. The intention and agreement phase are refined into offer specification, offer submission, offer analysis, and offer acceptance in line with the Montreal Taxonomy. The horizontal axis distinguishes the two negotiators' private areas and their shared communication. For simplicity, the process model does not take into account that one agent might be involved in several negotiations simultaneously or that parties might be non-monolithic.

A negotiation process starts with the information phase in which participants gather information on products, potential counterparties, and the socio-economic and legal environment. Once the parties finished collecting information, they advertise their willingness to negotiate. The intention phase starts with communication about the negotiation itself, i.e. with a meta-level negotiation. The parties define the issues of the negotiation, set up an agenda for the issues, and agree on a protocol to employ. This phase can either be an explicit part of the negotiation, or it can be implicit as the general rules for negotiating are given by the environment, by previous negotiations, by the parties' advertisements, or by any other institution. The meta-level negotiation can be revived during the overall negotiation.

Once the meta-level negotiation is finished for the first time, the parties enter into an offer exchange that is a back and forth in the intention and agreement phase: one of the agents privately specifies an offer and publicly submits it, i.e. she communicates it to the other agent. The receiver of the offer analysis it and decides on its acceptance with four possible outcomes: (1) the offer is accepted, (2) the negotiation is terminated, (3) the meta-level negotiation is revived, or, most commonly, (4) a counteroffer is specified and then submitted. Offer analysis can involve a new information phase.

In the offer specification and submission phases, an agent can not only propose and agreement but can as well shape arguments, reject the counterparty's offer without proposing a new one, withdraw an own offer, or any combination thereof. Obviously, these activities depend on the degrees of freedom granted by the agreed upon negotiation protocol.

Once an offer is accepted unconditionally, it becomes a contract. In the settlement phase, agents execute the contract, deliver goods, provision services, monitor compliance, etc. This general negotiation process model can be used to assess a negotiator's decision-making.

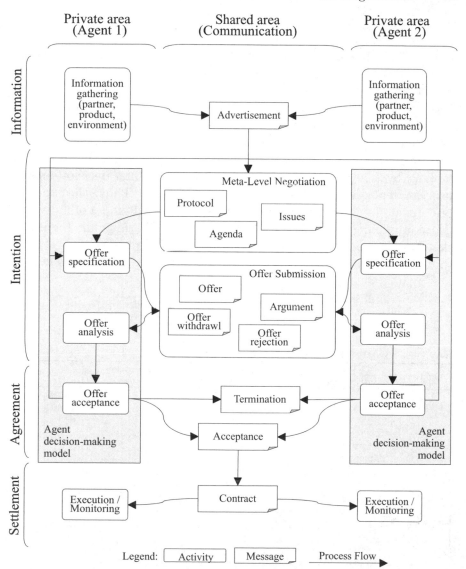

Fig. 3.2. Detailed process perspective on bilateral negotiations

Agent Decision Making Model

Figure 3.2 differentiates among private and shared information and activities. All messages passed from one agent to the other are shared information and the communication process is by its very nature a shared activity of sending and receiving messages. Important steps in a negotiation are, however, compiling and analyzing offers. Both activities are performed in private and based on a negotiator's cognition as well as private information, beliefs, and objectives. This individual decision-making by a negotiator in the overall non-individual decision-making process depends on the negotiator's preferences. Thus, the study of preferences in negotiations requires a more detailed model of individual decision-making than displayed in Figure 3.2. To this end, Figure 3.3 sketches the decision-making model more fine grained and in analogy to models from cognitive psychology (cf. McCabe, 2003, Fig. 1).

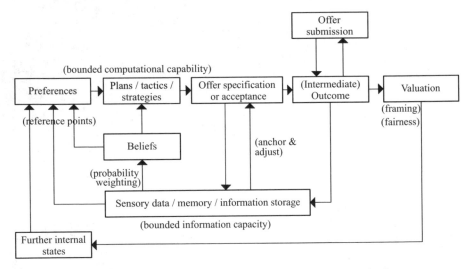

Fig. 3.3. Agent decision-making model in a negotiation

Assume a sub-process of the overall negotiation process depicted in Figure 3.2: an agent receives an offer from the counterparty, analysis it, decides on its acceptability in the agreement phase, returns to the intention phase to specify a new offer, and finally submits the new offer. From the agent's perspective, the incoming offer is an external event that changes the current status of the negotiation and determines an (intermediate) outcome. This outcome is—from a psychological perspective—internally processed in two ways. Firstly, it is a piece of information that enters the agent's memory and can be recalled later. Memory is an information storage device like a sheet of paper, a negotiation support system, the agent's brain, etc. Secondly, the outcome is evaluated, e.g. by comparing it to other possible outcomes and aspirations, and gives feedback to the agent's (further) internal states like, for example, her mood and other feelings. The term 'further' here indicates that other elements of the decision-making model, like preferences and beliefs, are internal states as well. However, as they are especially important in the present analysis, they are pinpointed whereas other internal states are summarized as 'further internal states'. From information in her memory, the agent forms new beliefs on her counterparty and the future negotiation process: What are the counterparty's preferences? How patient is the other agent? Which offer might just be acceptable? What is the likely outcome given the negotiation so far? The beliefs together with the agent's preferences enter a planning stage. A possible plan resulting from this is, for example, 'Make a new offer and ask for a better contract'. In the offer specification stage, finally, the agent compiles a counteroffer. For this, she can recall information from memory. Furthermore, preferences and beliefs enter the offer specification. The result is a new intermediate outcome, i.e. a new offer that is submitted to the counterparty. Other possible outcomes are offer rejection or offer acceptance.

3.1.3 Biases in Negotiations

The decision-making model outlined so far obviously is an abstraction of the real mental processes involved in a negotiation—it highlights the complexity of decision-making and points out different sub-processes for which negotiators might use heuristics and might fall prey to biases, i.e. systematic deviations from normative decision-making models like utility maximization.

In Figure 3.3, the words in parentheses exemplify which behavioral biases might occur along the offer analysis and specification process. The agent forms, for example, subjective beliefs based on objective information. This opens up the potential of subjective probability weighting (cf. Sec. 2.2). Furthermore, offer specification interacts with memory. If the agent compiles her offer by taking previous offers and adjusting them, then this allows for the emergence of the anchor and adjust bias—the agent is anchored at the old offer and adjustments towards the desired offer are usually insufficient (cf. Sec. 2.3.2).

All over the decision-making model, the boundedly rational agent is restricted by cognitive bounds and faces accuracy-effort trade-offs.

Memory, for example, usually is restricted: not storing all information decreases the accuracy of beliefs and thus decisions. On the other hand, selective information storage reduces the effort devoted to the task. Furthermore, combining preferences and beliefs and building a plan how to proceed is a computationally complex task and agents likely do not find an optimal plan but settle for a satisficing one.

The present study focuses on the systematic emergence of reference points and an attachment effect depending on offers exchanged in a negotiation. This will be detailed in Section 3.2. Before this, however, several other common biases in negotiations are briefly outlined in the following.[7] Some of them are general biases in individual decision-making that apply—among other decision contexts—in negotiations:

Anchoring and adjustment: In several studies it has been found that agents frequently start the estimation of an unknown value by taking any available information (whether relevant or not) as initial anchor and then estimate the unknown value by adjusting this anchor. The last digits of a social security number, for example, can be used to experimentally demonstrate how subjects incorporate objectively irrelevant information in their estimation of valuations (cf. Sec. 2.3.2). In negotiations, the initial offer of one negotiator might anchor the counterparty on this offer and any subsequent adjustment away from it usually is insufficient. If, for example, a seller starts with a high initial price, the final price the negotiators agree upon will be higher than if the seller would have started with a moderate price. Evidence for the effect of anchors in single-issue negotiations has been reported repeatedly Northcraft and Neale (1987), Kahneman (1992), Thompson (1995), Ritov (1996), Whyte and Sebenius (1997), and Kristensen and Gärling (1997b, 2000). As mentioned before, anchoring and adjustment relates to the interplay of offer specification and memory (cf. Figure 3.3).

Framing: Oftentimes, one and the same choice situation can either be framed positively (e.g. as a gain) or negatively (e.g. as a loss). It has frequently been reported that the framing has an impact on agents' attitude towards risk: they tend to be risk-seeking for losses and risk-averse for gains (Sec. 2.1.2). In the context of negotiations, it has been found that parties tend to make stronger concessions to the counterparty when the overall negotiation is framed positively rather than negatively Bazerman, Magliozzi, and Neale (1985), Bot-

---

[7] Comparable collections of biases in negotiations are provided by Neale and Bazerman (1991, Ch. 3 & 4), Bazerman and Neale (1992, Part I), Bazerman, Curhan, Moore, and Valley (2000), and Bazerman (2006, Ch. 10).

tom and Studt (1993), Olekalns (1997), and De Dreu and McCusker (1997). In the decision-making model presented before, framing influences the valuation of offers.

Availability: According to the availability bias, decision-makers tend to overestimate the probability of unlikely events if instances of that event are easily available in their memory (Tverksy and Kahneman, 1974). Not all past experiences of a negotiator might be coded equally in memory—some are easier to retrieve (they are better available), and hence, their likelihood is overestimated. Opportunity costs, for example, can be seen as less concrete than out-of-pocket costs and empirical evidence has been found that opportunity costs are less likely to be included in decision-making during negotiations (Northcraft and Neale, 1986; Neale and Bazerman, 1991, Ch. 3). Further evidence for the availability bias in negotiations is presented by Neale (1984) and Pinkley, Griffith, and Northcraft (1995). The availability bias effects the formation of probabilities and beliefs in negotiators' decision-making.

Overconfidence: Several studies found evidence that people tend to overestimate their abilities and decision-makers tend to be overly confident in the correctness of their decisions (e.g. Fischhoff, 1982). In negotiations, overconfidence leads to excessively optimistic judgments about the likelihood of getting a good outcome (Bazerman and Neale, 1982; Lim, 1997). Kramer, Newton, and Pommerenke (1993), for example, found that 68% of their students predicted that the outcomes they would negotiate would fall in the upper 25% percent of the outcomes negotiated by their fellow students. This obviously is an overly optimistic prediction. An alternative name for this bias is *self-enhancement bias*. As the availability bias, overconfidence is related to a negotiator's beliefs.

Besides these aforementioned biases, Neale and Bazerman (1991, Ch. 3), point out common mistakes in evaluating the law of small numbers, the confirmatory evidence bias, and judgment of causation as further effects frequently reported in psychological studies. However, their application in negotiations is not (yet) supported empirically. Furthermore, many negotiators are prone to the following cognitive and behavioral patterns that are specific for negotiations:

Fixed pie illusion: Many negotiators disregard the integrative potential of multi-issue negotiations, focus on competitive issues, and assume to play a constant-sum game. As a result, agreements are oftentimes inefficient or the negotiation is terminated without agreement (Bazerman, Magliozzi, and Neale, 1985; Thompson and Hastie, 1990; Thompson and DeHarpport, 1994; Fukuno and Ohbuchi, 1997).

Illusion of conflict: Negotiators oftentimes falsely assume that their own preferences are in opposition to their opponent's preferences. Hence, they see a compromise which is good for the counterparty as bad for themselves (Thompson, 1990; Thompson and Hastie, 1990; Thompson and Hrebec, 1996). *Incompatibility bias* is an alternative name for this effect; it is closely related to the fixed pie illusion and both affect a negotiator's beliefs.

Reactive devaluations: The phenomenon of reactive devaluations of the counterparty's offers directly follows from the illusion of conflict. Parties devalue any proposal made by the counterparty just because it originates from the counterparty and they assume that it cannot be beneficial for them (Ross and Stillinger, 1991). This bias relates to the valuation of incoming offers in the decision-making model.

Escalation of conflict: Negotiators sometimes tend to escalate a conflict even if terminating the negotiation or giving in to the counterparty's demands would be beneficial. This can, for example, frequently be observed in labor disputes when—under pressure from the public opinion that increases the effect—neither the union wants to end a strike nor the employers want to meet the increase in salary demanded by the union. Among the hypothesized reasons for such an escalation are cognitive dissonance theory that requires to rationalize one's own previous choices (Festinger, 1957) and the avoidance of realizing a loss rather than postponing the decision (Bazerman and Neale, 1983; Bizman and Hoffman, 1993; Diekmann et al., 1996). Escalation of conflict, cognitive dissonance, and commitment to prior actions affect a negotiator's further internal states.

Ignorance of the other's behavior: In game theory, the focal element of Nash equilibria is that all players' equilibrium strategies are best responses to the others' equilibrium strategies. For finding such equilibrium strategies it is inevitable to consider the others' strategic considerations. Yet, in real life negotiation decision-making the strategic character is oftentimes disregarded and negotiators do, for example, frequently not take into account that an acceptance of their offer by the counterparty ex-post signals that the offer might have been to generous, i.e. they fall prey to a variant of the winner's curse (Bazerman and Carroll, 1987; Carroll, Bazerman, and Maury, 1988; Valley, Moag, and Bazerman, 1998). The insufficient reasoning on the counterparty's likely behavior relates to bounded capabilities in strategic reasoning: in some experiments it was found that most individuals have the analytical capability to understand the winner's curse if it is presented to them; they do, however, not find their 'best' strategy without assistance (Neale and Bazerman, 1991, Ch. 4).

Egocentrism: Judgment of fairness of an agreement oftentimes differs between parties in a negotiation. Each party tends to be egocentric and build its own subjective notion of fairness. The different egocentric fairness perceptions let parties dismiss outcomes which are identified as fair by the counterparty (Camerer and Loewenstein, 1993; Loewenstein, Issacharoff, and Camerer, 1993; Babcock et al., 1995). Fairness perceptions relate to the valuation in a negotiator's decision-making model.

Summary

Research on negotiations has brought up several systematic biases that influence decision-making in negotiations. The different effects do not contradict each other; they rather apply independently or, like the fixed pie illusion and the illusion of conflict, are closely related to each other. However, to date there is no coherent 'theory of negotiation analysis' but rather a collection of self-sufficient effects. Bazerman et al. (2000, p. 283)—who authored and co-authored many of the articles cited above—comment this line of research with the sentence: 'Clearly, a goal to provide useful information that could lead to the debiasing of negotiators guided this research.' The objective of finding biases is not (as skeptics sometimes assume) to prove how naive people are; it rather is to help them in overcoming biases and make the decisions they would want to make would they know about the bias—in the absence of perfect rationality by all negotiators at all times, description of real behavior is seen as necessary preparation for prescription. In Section 3.2 an additional bias is added to the list of common biases in negotiations, namely the *attachment effect*. The effect relates to the decision-making model sketched in Figure 3.3 as follows: Offers (either from the counterparty or from the negotiator herself) enter memory, this affects beliefs and expectations which in turn influences preferences via a potential change of the reference point. Preferences then enter into subsequent choice. This attachment bias adds to the descriptive part of negotiation analysis and can be exploited in preparing to negotiate rationally.

3.2 Origin of Reference Points

In preparing for a negotiation, each party should sort out its preferences. Negotiators are continually asked to evaluate whether they prefer one alternative to the other during a negotiation. Moreover, they oftentimes have to make trade-offs and have to specify what they are willing to give up for claiming more on another issue (Raiffa, 1982, Ch. 11). Essentially, preferences are the basis for decision-making in negotiations. The implications of reference-dependent preferences have been studied extensively over the last decades. The origin of reference points, on the contrary, is a grossly understudied topic. The origin and endogenous change of reference points in negotiations is considered in more detail in the following.

Behavioral economics offers several insights in the psychological context of consumer decision-making. Loss aversion and the implications of the shift of reference points were, for example, addressed in Section 2.2. In multi-issue choice, the shift of the reference point on one issue changes the slope of indifference curves, and hence, the trade-offs across issues. A question not addressed so far is which events cause a shift of reference points? In literature, two different causes are discussed: assignment of *property rights* is the traditional explanation and *expectations* in future property rights and consumption is a more recent one.

3.2.1 Property Rights

Property rights are relationships of agents (i.e. individuals, corporations, nations, etc.) and objects. They assure that an agent possesses the object and legal systems usually assign the right to control use over one's property and to transfer the right to other agents. In a consumer good market, for example, sellers transfer the possession of a good to buyers in exchange for money. In behavioral economics, the status quo of property rights is oftentimes assumed to be the agents' reference point.

A frequent empirical finding is that consumers who posses a good demand a higher price for selling the good, than they would be willing to pay wouldn't they own the good. By experimentation, economists and psychologists were able to show that this discrepancy—or willingness-to-pay/willingness-to-accept gap (*WTA-WTP gap* for short) as it is oftentimes termed—is not due to strategic considerations of market participants but due to a dependence of preferences on property rights.

The increased value of a good to an individual when the individual obtains property rights is termed *endowment effect* (Thaler, 1980). This effect manifests the aversion to losses inherent in prospect theory and contrasts the Coase theorem which asserts that the allocation of resources will be independent of property rights when costless trades are possible (Coase, 1960). The Coase theorem relies on the assumption that property rights do not affect valuations; however, with an endowment effect, valuations are affected and agents request a higher price for giving up a good, than they are willing to pay in order to acquire the same good (Kahneman, Knetsch, and Thaler, 1990; Tversky and Kahneman, 1991).

Classical Experimental Evidence

The classical experiment to test for an endowment effect is as follows: Coffee mugs or other goods are given by random to half of the subjects (e.g. students in a classroom) and then the willingness to sell the mug is elicited from the students who got one by chance. Furthermore, the willingness to pay is elicited from students who were not endowed with a mug. The elicitation can be performed with questionnaires inducing incentive compatible behavior. Typical results are that students holding a mug in their hands ask for more than twice the price that students without a mug are willing to pay.

The probably most commonly cited study using this design with various different goods is reported by Kahneman, Knetsch, and Thaler (1990). They invited subjects to a laboratory and randomly allocated coffee mugs to half of them. Subjects with a mug were potential sellers; subjects without a mug were potential buyers. Conventional analysis predicts that one half of the products should be traded. If value is unaffected by ownership, the distribution of valuations is the same for sellers and buyers. About half of the sellers will have a lower valuation than about half of the buyers and trades will be beneficial for them. Contrary, if there is an endowment effect, the median seller valuation is higher than the median buyer valuation and less than half of the coffee mugs will be traded. Therefore, trade volume is a simple measurement to test for endowment effects.[8] Besides coffee mugs, Kahneman et al. used pens, binoculars, and chocolate bars to demonstrate the endowment effect.

---

[8] Depending on the market mechanism, strategic considerations might enter the preference elicitation. If, however, the price is determined by random and, thus, all subjects are price takers, strategic considerations are ruled out.

Kahneman et al. expect that for goods that are purchased for resale instead of consumption there is no endowment effect. Goods in induced-value experiments are examples for goods held solely for reselling them to the experimenter at the end of the experiment. The authors used this as benchmark for their experiment. Factors like standard bargaining habits, transaction costs, or misunderstanding could influence trade volume in the experiment. However, if the conventional prediction of half the products being traded holds for induced-value markets, but not for consumption goods, endowment effects seem to be a plausible explanation.

The data gathered by Kahneman et al. shows that markets for induced-value goods and consumption goods yield sharply different results. In their experiment 1, for example, they had 44 subjects per session. Half of them were potential buyers and half of them potential sellers which leads to the conventional prediction of 11 trades per session. With induced values there were 10 to 12 trades and prices equaled expected prices; with consumption goods there were 1 to 5 trades and median buyer reservation prices for coffee mugs were about $ 2.25, whereas median seller reservation prices were $ 5.25. Kahneman et al. interpret the low trade volume and large difference in the willingness to pay and willingness to accept as clear evidence for an endowment effect for consumption goods.

The results hold with other consumption goods besides coffee mugs, repeated experiments with the same subjects, and exchanging two consumption goods instead of one good against money.

Evidence from the Field

The WTA-WTP gap is not a phenomenon of contrived experiments with students and inexpensive goods, it rather is an effect found in the field for many different classes of goods, subject pools, and elicitation procedures. Bishop and Heberlein (1979) and Brookshire, Randall, and Stoll (1980), for example, found evidence for the discrepancy in studies on the valuation of hunting licenses, and Brookshire and Coursey (1987) in a study on the density of park trees. Furthermore, Camerer (2001) reviews field evidence for loss aversion in stock markets, labor economics, consumer goods, and purchase of insurances and Samuelson and Zeckhauser (1981)—who introduced the term *status quo bias* for the tendency that decision-makers oftentimes favor the current situation over any change—report data suggesting that individuals use the status quo as reference point with respect to, e.g., financial investments and choice of medical plans.

In a review of more than 40 studies on the WTA-WTP discrepancy, Horowitz and McConnell (2002) find that the discrepancy is highest for non-market goods, next highest for ordinary private goods, and lowest for experiments involving forms of money. This is in line with the notion of preference construction and stabilization: the less familiar it is to express preferences over a good in terms of money, the more likely it is that preferences are not simply revealed, and hence, preferences are more prone to being influenced by the endowment effect. However, not all authors are so enthusiastic about the accumulated evidence; see Plott and Zeiler (2005) for a recent critical discussion of literature on the endowment effect. The authors argue that results reported in literature might to a large extend be experimental artifacts, especially due to strategic considerations that might arise from misconceptions of incentive compatible elicitation procedures and a lack of training with the elicitation procedures used.

Status Quo Bias with Multiple Issues

Most empirical evidence concerning the effect of property rights, loss aversion, and a status quo bias study decision-making with respect to a single issue. Among the few exceptions studying the status quo in multi-issue decisions is the field study by Hardie, Johnson, and Fader (1993). They analyze the brand choice by consumers and take the brand chosen in the last purchase as the status quo. With this, they find clear evidence that consumers evaluate their purchase relative to the status quo and exhibit loss aversion on single issues. In another study, Tversky and Kahneman (1991, cf. Sec. 2.2.2) report further evidence for loss aversion concerning the status quo defined along several issues.

Extensions and Limitations

The classical interpretation of the endowment effect is that current property rights immediately affect an individual's reference point. Two extensions thereof are the *history of ownership* effect and the *source effect*. Strahilevitz and Loewenstein (1998) found that the endowment effect gradually becomes stronger with duration of ownership and remains (though less strongly) even after property rights for a good were transferred to someone else. They conclude that there is a history of ownership effect. Furthermore, the strength of the endowment effect depends on the source of the good, i.e. whether one obtains ownership by change or because of exceptional performance in a task: Loewenstein and Issacharoff (1994) report that subjects who 'earned' a good are more attached to it than subjects who received it by chance.

Two limitations of the applicability of the endowment effect are re-sale goods and experience. Kahneman, Knetsch, and Thaler (1990) and many other authors report that they do not find evidence for an endowment effect for abstract, induced-value tokens or for cash. Furthermore, List (2003) reports that experience with a good lets individual behavior converge to the rational choice prediction that property rights do not influence preferences. However, to date there is no well-developed economic theory on learning and the empirical evidence is inconclusive. For decisions that occur infrequently like marriage, negotiating an employment contract, or deciding on a health plan the effect of learning and experience has natural limitations. Frequently repeated decisions like for example in stock markets, on the other hand, are more likely to be influenced by learning (Starmer, 2000). Consequently, the descriptive validity of different models is likely to depend on the domain where it is applied.

3.2.2 Expectations

While the status quo of property rights is the traditional explanation, expectations in either future property rights or in future consumption have been discussed more recently as potential cause of reference points. The role of expectations is most explicitly studied by Köszegi and Rabin (2006). It was, however, already noted earlier, e.g., by Tversky and Kahneman (1991, pp. 1046–1047) who state that 'although the reference state usually corresponds to the decision-maker's current position, it can also be influenced by aspiration, expectations, norms, and social comparisons.' Köszegi and Rabin present a model of reference-dependent preferences that does to a large extend draw on traditional prospect theory with respect to the implications of a reference point, i.e. mainly loss aversion. Their model does, however, formalize how the probabilistic beliefs of a consumer held in the recent past about future states influence her multi-issue reference point. The authors explicitly note that this different interpretation of the origin of reference points is not in contrast to the vast empirical evidence on the endowment effect. They rather argue that in almost all these studies the status quo can reasonably be expected to be the future status quo and, thus, their model's prediction coincides with the status quo bias for consumption goods. Beyond that, expectations instead of property rights as cause of reference points are more widely applicable: they allow to explain auction fever, the role of experience, the absence of an endowment effect for professional dealers, its absence for cash, loss aversion for the participation in future events, employees' aversion to wage cuts, etc.

Auction Fever

Auction fever can roughly be defined as bidders outbidding their initial reservation price in an auction where this cannot be explained by information gathered during the auction (Ku, 2000). Others define it as the 'emotionally charged and frantic behavior of auction participants that can result in overbidding' (Ku, Malhotra, and Murnighan, 2005, p. 90). Four potential causes of auction fever have been identified so far: competitive arousal, escalation of commitment, pseudo-endowment, and an attachment effect. Competitive arousal means that the presence of competitors, time pressure, etc. get a bidder so excited that he does not stop bidding when the price exceeds her valuation for the good at auction. The escalation of commitment hypothesis assumes that bidders continue bidding to justify their prior bids. Both explanations are related to common biases in negotiations (cf. Sec. 3.1.3).

The pseudo-endowment effect is a variant of the traditional endowment effect: it is assumed that bidders perceive ownership of an objectively un-owned item (Ariely and Simonson, 2003; Heyman, Orhun, and Ariely, 2004). The reasoning is that the high bidder in an iterative auction might feel an entitlement to win the auction and already integrates the good that is being auctioned in her 'psychological endowment' before the auction is won. Explaining auction fever via psychological endowment and the bidders' misconception of the auction mechanism is rather complicated compared to an explanation via expectations and attachment. Being high bidder for an extended period likely increases a bidder's subjective belief of winning the auction and thus—according to the model by Köszegi and Rabin (2006)—her reference point for the good. With loss aversion, this shift of the reference point increases the utility difference between winning and loosing the auction, and hence, the bidder is willing to submit higher bids than she would have submitted at the beginning of the auction process. To date, empirical studies on auction fever do not allow discriminating pseudo-endowment and expectation-based attachment (Abele, Ehrhart, and Ott, 2006).

Experience, Dealers, and Cash

Several studies have attempted to find the boundaries of the endowment effect. It has widely been noted that the effect does not occur for cash, abstract induced-value tokens in experiments, or resale goods. Furthermore, experience lets the endowment effect diminish and professional dealers are not prone to the effect (e.g. Kahneman, Knetsch, and Thaler, 1990; List, 2003, 2004; Novemsky and Kahneman, 2005). Given these limitations and status quo property rights as reference points, it is difficult to a-priori predict whether a specific individual will show loss aversion for a specific good: Is this person more like a regular coffee drinker consuming the coffee mug or more like a dealer intending to sell it? And is her accumulated experience enough to rule out loss aversion?

Again, an argumentation via expectations is way easier: A dealer buys a good for selling it. Thus, her reasonable (or even rational) expectations are not to possess the good in the future. Hence, she does not perceive a loss when selling the good. A subject acquiring the same good in an experiment might, on the other hand, expect to keep it and, thus, become attached to it. The same argument works for cash: cash is an exchange medium and, thus, hardly anybody who gets cash from an ATM expects to possess the same bills for an extended period. Consequently, there is no loss aversion.

The role of experience can easily be explained via expectations as well. Experienced collectors of sports cards, for example, know that exchanging cards is part of being a collector. Thus, they do not expect to keep a card forever and do not exhibit loss aversion. Inexperienced collectors, on the other hand, might expect a low probability of parting with a specific card and, thus, perceive a loss when giving it away.

Summary

Most studies on reference-dependent preferences assume that status quo property rights serve as reference point. Virtually all empirical evidence reported in this vein can as well be explained by an expectations-based approach as presented by Köszegi and Rabin (2006). Furthermore, expectations allow explaining boundaries of loss aversion and the endowment effect relatively easy. Both origins of reference points might be present in negotiations: negotiators might already account for psychological endowment during a negotiation or their expectations in the likely outcome and, thus, their future property rights and consumption might change. Anyways, both approaches favor the assumption that preferences are endogenous to market processes. This is modeled for negotiations in the following.

3.3 The Attachment Effect in Negotiations

Analyzing the shift of reference points during a negotiation via a negotiator's expectations is more intuitive than via her psychological endowment. Thus, the following presentation focuses on the influence of offers on beliefs rather than pseudo-endowment. It is, however, note-worthy that the same observable behavior might be caused by pseudo-endowment. Following Köszegi and Rabin (2006), the effect of offers on preferences is termed *attachment effect*. The effect can be defined as an increased attachment to issue-specific outcomes of a negotiation as result of the exchange of offers, expectations in the outcome, reference-dependent evaluation of offers and agreements, and loss aversion.

3.3.1 Graphical Example[9]

To illustrate the attachment effect, Figure 3.4 shows a bilateral two-issue alternating-offer negotiation in a sequence of Edgeworth-boxes. The two negotiators are labeled S (she) and H (he) and the two issues are denoted as A and B. The issues are normalized to the unit interval and either party prefers more to less on each issue. Negotiator S's share is measured from the lower left corner of the box. Accordingly, the share of H is measured from the upper right corner. For simplicity, only negotiator S is prone to an attachment effect in this example.

Figure 3.4a

Figure 3.4a shows the initial, exogenously given setup at time $t = 0$ (time is indicated by superscripts): Two of S's indifference curves are displayed and S has a reference point r^0 which might initially be at zero on both issues.

Figure 3.4b

At $t = 1$, H makes the initial offer x_H^1 in the negotiation. This offer influences S's beliefs about the likely outcome. Would H have offered to settle for $x_H^{1'} = \langle 1, 1 \rangle$, for example, S might have accepted this of-fer right away without further delay as it is her ideal outcome. If H would have offered $x_H^{1''} = \langle 0, 1 \rangle$, as another example, this might have been interpreted as signal that a high value on issue B is easy to agree on. But instead of $x_H^{1'}$ or $x_H^{1''}$ (or any other of infinitely many alterna-tives) H offered x_H^1 and this choice affects S's expectations in the final agreement. The attachment effect applies: Expectations influence S's reference point which shifts from r^0 to r^1. The shifting reference point in turn implies that S's perspective on the agreement space changes, i.e. her indifference curves change as trade-offs are affected by the eval-uation as gains or losses (cf. Sec. 2.2.2, esp. Fig. 2.7). New indifference

---

[9] A related example of shifting reference points in negotiations is given by Gimpel (2007).

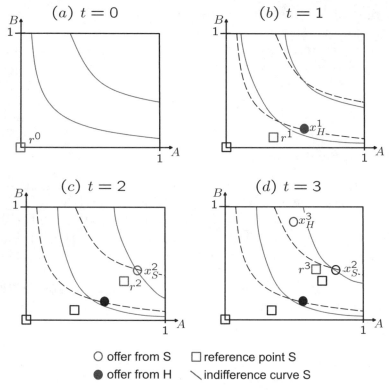

Fig. 3.4. Example of the attachment effect in a negotiation

curves are plotted as solid lines, the a-priori indifference curves as dotted lines.

Figure 3.4c

At $t = 2$, S rejects the offer by H and makes a counteroffer x_S^2. Like H's offer, her own offer influences her beliefs in the likely agreement of the negotiation. Again, her reference point shifts: it becomes r^2.

Figure 3.4d

Finally, at $t = 3$, it is H's term again: he makes a new offer x_H^3, the specific offer influences S's expectations and, thus, her reference point which becomes r^3.

At this point, a comparison of S's indifference curves at $t = 0$ (dotted lines) and at $t = 3$ (solid lines) suggests that the reference point has an effect on S's behavior. With her a-priori preferences, S would perceive the offer by H as quite generous—x_H^3 gives her higher utility than her own last offer x_S^2. Thus, given the a-priori preferences, she might be expected to agree on x_H^3. With her current preferences at $t = 3$, however, H's offer is evaluated differently. The negotiation process has changed her reference point and trade-offs. Given the current reference point r^3, x_H^3 is not as good as S's last offer. From r^3, her offer x_S^2 would be a gain on issue A and neutral on issue B. Accepting x_H^3, however, means perceiving a loss on issue A which cannot be compensated by the gain on B. Thus, it might very well be that S rejects x_H^3 and the negotiation continues.

Discussion

A few things about this example are noteworthy: Firstly, and most importantly, the reference point and the indifference curves do not change arbitrarily. In all instances, the reference point is anchored at its location from the previous period and adjusted towards the current offer. The intuition is that if an offer is reflected in a negotiator's expectations, then in a way that the negotiator assumes this offer to become more likely as agreement. Or, in other words, if H offers S more on issue A than she expected to obtain up to now, than her expectation in the outcome on issue A will not be lowered. It will, however, not exceed the offer either. The change of indifference curves based on a reference point is qualitatively as proposed by Tversky and Kahneman (1991).

Secondly, the reference point changes issue-wise, i.e. it is not that S's aspiration in the utility she will likely obtain from an agreement changes but her reference point is defined on each issue individually. This is in line with the models proposed by Tversky and Kahneman (1991) and Köszegi and Rabin (2006). Thirdly, it is noteworthy that the adaptation of expectations might differ between issues. Just as the endowment effect was found to differ between types of goods, the attachment effect might differ between issues in a negotiation. Fourthly and finally, the example is not meant to illustrate how strongly the attachment effect applies. Figure 3.4 exemplifies the qualitative implications of an attachment effect in multi-issue negotiations. Whether real negotiators are prone to this bias and how strongly it affects them is a matter of empirical studies. To this end, the attachment effect is formalized in the next section and then tested and quantified in two experiments.

3.3.2 Formalization

The attachment effect that was exemplified so far can be modeled as follows: Two parties negotiate to reach an agreement on $K \geq 1$

issues simultaneously. Parties take turns in exchanging offers at discrete points in time $t = 1, 2, 3, \cdots$, the agreement space is given as $X = [\underline{x}_1, \overline{x}_1] \times [\underline{x}_2, \overline{x}_2] \times \cdots \times [\underline{x}_K, \overline{x}_K] \subseteq \mathbb{R}^K$. Each offer $x^t = \langle x_1^t, x_2^t, \cdots, x_K^t \rangle \in X$ is an element of the agreement space. In the following, preferences and reference points of only one of the two negotiator's—the so called *focal negotiator*—are considered for ease of notation. The exact same consideration can be applied to her counterparty. It is assumed that the focal negotiator has monotone preferences, i.e. w.l.o.g. $\forall\, k \in \{1, 2, \cdots, K\} : x_k \succeq y_k \Leftrightarrow x_k \geq y_k$. Furthermore, the negotiator has a time-dependent reference point $r^t = \langle r_1^t, r_2^t, \cdots, r_K^t \rangle \in X$ in the agreement space. The reference point could as well be modeled to depend on the history of offers; however, as exactly one offer is made at any point in time, offer-dependence and time-dependence are equivalent in this model.

Let the a-priori reference point of the focal negotiator before the negotiation starts be $r^0 = \langle r_1^0, r_2^0, \cdots, r_K^0 \rangle \in X$. Then the attachment effect implies that the reference point is updated each time the negotiator updates her expectations. This happens at each point in time during the negotiation when an offer is made by either party, i.e. at $t = 1, 2, 3, \cdots$. The issue-wise update is defined recursively as follows $\forall\, k \in \{1, 2, \cdots, K\}$:

$$
r_k^t = \begin{cases}
r_k^{t-1} + f_k^+(x_k^t - r_k^{t-1}) & \text{if } t = 1, 3, \cdots \text{ and } x_k^t \geq r_k^{t-1}, \\
r_k^{t-1} - f_k^-(r_k^{t-1} - x_k^t) & \text{if } t = 1, 3, \cdots \text{ and } x_k^t < r_k^{t-1}, \\
r_k^{t-1} + g_k^+(x_k^t - r_k^{t-1}) & \text{if } t = 2, 4, \cdots \text{ and } x_k^t \geq r_k^{t-1}, \\
r_k^{t-1} - g_k^-(r_k^{t-1} - x_k^t) & \text{otherwise}
\end{cases}
$$

where $f_k^+(\cdot)$, $f_k^-(\cdot)$, $g_k^+(\cdot)$, and $g_k^-(\cdot)$ are issue-specific update functions for the reference point. Note that the negotiators take turns in making offers x^t. W.l.o.g. the focal negotiator's counterparty starts with an initial offer x^1 and decides on $x^3, x^5, \cdots$ whereas the focal negotiator herself proposes agreements at even times t. Then the functions $f_k^+(\cdot)$ and $f_k^-(\cdot)$ capture the influence of the counterparty's offers on the focal negotiator's reference point and $g_k^+(\cdot)$ and $g_k^-(\cdot)$ the influence of her own offers.

According to the above definition, the function $f_k^+ : \mathbb{R}_+ \to \mathbb{R}_+$ defines the increase of the reference point on issue k in case the focal negotiator receives an offer of x_k^t for issue k from her counterparty and this offer grants a higher value than the reference point was before receiving the offer. In the example sketched in Figure 3.4 the first offer by H on issue A is a situation where such an update would apply. The function is assumed to have the following characteristics:

- If the reference point and the offer coincide, the reference point is not changed (lower bound: $f_k^+(0) = 0$).
- A greater difference of reference point and offer does not lead to a smaller update (monotonicity: $\forall\, a, b \in \mathbb{R}_+ : f_k^+(a + b) \geq f_k^+(a)$).
- The offer received is the limit for the update of the reference point (upper bound: $\forall\, a \in \mathbb{R}_+ : f_k^+(a) \leq a$).

The lower bound and monotonicity together imply $f_k^+(\cdot) \geq 0$ which means that an offer higher than the reference point will not change the negotiator's expectations in a way that the reference point diminishes. The same characteristics are assumed for the other update functions $f_k^-(\cdot)$, $g_k^+(\cdot)$, $g_k^-(\cdot)$ that capture the effect of offers on the negotiator's reference point when either the offer is lower than the reference point or the offer comes from the negotiator herself.

The partial adaptation of a new offer as reference point is in line with the finding that changes of reference points are not all or nothing but are a slowly progressing process (Strahilevitz and Loewenstein, 1998). Furthermore, the issue-specific update functions take into account that the endowment effect was found to be rather strong for consumption goods and virtually non-existent for resale goods and money (e.g. Kahneman, Knetsch, and Thaler, 1990). Finally, the fact that update functions might differ across negotiators integrates the fact that it is an individual effect that is influenced by experience or other personal characteristics (e.g. List, 2003).

Extensions

In the present formalization, changes of a negotiator's reference point solely depend on its prior location and the current offer. Additionally, one could expect that the entire history of offers has an influence on the negotiator's expectations. Furthermore, the update functions themselves might be time-dependent. However, for simplicity neither of these extensions is considered here.

Existing Experimental Evidence

In the review of related work at the beginning of this study (Sec. 1.1), the experiments by Kristensen and Gärling (1997a) and Curhan, Neale, and Ross (2004) were reviewed. Both provide evidence that the process of negotiating can affect reference points. Kristensen and Gärling report that their subjects who play the role of buyers of condominiums oftentimes adopt the sellers' initial offers as reference points. The authors reason that a seller's initial offer may provide information about the regular market price and buyers might thus perceive a gain when the final price is below the initial offer. This argumentation is in line with the attachment effect model; after reversing the direction of monotonicity of preferences (buyers prefer lower prices whereas above it was assumed w.l.o.g. that the negotiator prefers higher values) and dropping issue-specific indices due to the restriction on one issue ($K = 1$), the results by Kristensen and Gärling suggest $\exists\, a \in \mathbb{R}_+ : f^-(a) > 0$. Verbally this means that a counterparty's unfavorable offer, i.e. an offer by the seller that is higher than the negotiator's expectation, influences the negotiator's reference point as she expects to get a less desirable agreement.

Based on a supposition by Kahneman (1992), Kristensen and Gärling (1997a) assume that the adoption of a reference point is all-or-none. Thus, in terms of the attachment effect model this rather restrictive assumption is $\forall\, a \in \mathbb{R}_+ : f^-(a) = a$. The data the authors present is—as any experimental data—noisy and does not allow to precisely conclude the exact functional form. It just allows to reject $\forall\, a \in \mathbb{R}_+ : f^-(a) = 0$. Under their premise of all-or-none adoption, Kristensen and Gärling (1997a) conclude $f^-(a) = a$. While this is in line with the attachment effect model, it is only a special case and the model allows for more variety in the update functions. Another interesting aspect about the results presented by Kristensen and Gärling arises from their experiment number 4. They told their subjects to imagine being tourist in a foreign country and that in this particular country bargaining about prices is very common. The experimenters' assumption was that this information reduces the effect of initial offers as reference points: If bargaining is common then the initial price demanded by a seller likely is exorbitantly high and does not signal much about the final price. The data reported on the experiment indeed suggests that initial offers had no significant effect under this information condition. This effect of exorbitant offers might be modeled by concavity of $f^+(\cdot)$ and especially $f^-(\cdot)$. If the difference of reference point and offer is rather small and the offer thus appears reasonable, its impact on the reference point is relatively strong. With increasing difference of offer and reference point, on the other hand, marginal sensitivity diminishes (the functions are concave) and an unrealistic, exorbitant offer only affects the reference point relatively weakly.[10]

A second experiment on the endogenous change of preferences in negotiations is presented by Curhan, Neale, and Ross (2004); it was as well introduced in Section 1.1. The authors report evidence that their subjects' preferences are influenced by the offers exchanged in a multi-issue negotiation. Subjects tended to express higher preferences for contracts once they had offered them. This tendency was even stronger when a contract became the final agreement. Curhan et al. present these results as evidence for cognitive dissonance in negotiations. The results reported seem, however, to be in line with the attachment effect as well. If the offer a negotiator makes lets her reference point move towards this offer then this changes her preferences. Assume, for example, that the offer is perceived as gain on all issues from the a-priori reference point. After making the offer, the reference point will be closer to the offer but the offer will nevertheless be evaluated as a gain. Other offers that initially lay on the same indifference curve but require extreme trade-offs and have rather low values on one or some issues are partially evaluated as loss from the new reference point. Thus, they no longer lie on the same indifference curve as the offer but are perceived as less desirable. The effect would be even stronger for the final agreement as this influences expectations, and hence, reference points more strongly than mere offers. This reasoning might (partially) explain the changes of preferences Curhan et al. attribute to dissonance theory. Whether this really is the case can, however, not be judged based on the data reported by Curhan et al.

Anyways, the discussion of the work by Curhan et al. is not meant to refute the relevance of dissonance theory for endogenous preference changes in negotiations. It rather suggests that there might be two effects—dissonance and attachment—that potentially cannot yet be disentangled precisely.

3.3.3 Simplifications of the Model

The above formalization of the attachment effect is rather general without assuming specific functional forms for the shift of a reference point. In the following, simplified versions of the model are presented.

---

[10] To even further strengthen this effect it would be necessary to relax the monotonicity assumption for update functions. As the remainder of the present work does not deal with unrealistic, exorbitant offers this is neglected for simplicity.

Rational Choice

The standard rational choice model of exogenously given and fix preferences is a special case of the attachment effect model with $\forall\, a \in \mathbb{R}_+$: $f^+(a) = 0$, $f^-(a) = 0$, $g^+(a) = 0$, and $g^-(a) = 0$. The existence of a reference point is still possible; it does, however, not change over time.

Single-Issue

Several studies have analyzed reference-dependent preferences in single-issue bargaining, i.e. for $K = 1$. Game theoretic models are presented by Shalev (2002), Li (2004), Hyndman (2005), and Compte and Jehiel (2006). In the model by Compte and Jehiel, for example, players are assumed to evaluate outcomes relative to a reference point that is equivalent to the most generous offer the counterparty has made so far. This setup can be reflected by the attachment effect model with $\forall\, a \in \mathbb{R}_+$: $f^+(a) = a$ and $f^-(a) = 0$, $g^+(a) = 0$, $g^-(a) = 0$.[11] Compte and Jehiel add costs of delaying an agreement and a structure of bargaining phases that consist of several offers each and have an exogenous breakdown probability after each offer. With this, the authors find a subgame perfect equilibrium for the game and in equilibrium there are inefficiencies and gradualism, i.e. unlike in the Rubinstein game, parties to not reach an agreement with the first offer but exchange several offers before agreeing. At the end of their analysis, Compte and Jehiel (2006) note that their assumption of $f^+(a) = a$ might be relaxed to $f^+(a) = \beta\, a$ where $\beta \in (0, 1)$ would measure the sensitivity of the reference point to prior offers.[12] The cases $\beta = 0$ and $\beta = 1$ would relate to a standard rational choice model and the model studied by the authors, respectively. This extension to linear updates makes the analysis—according to Compte and Jehiel—more complicated without changing the qualitative insights. The model by Li (2004) is similar to the one by Compte and Jehiel (2006) and exhibits gradual bargaining in equilibrium as well. Hyndman (2005) on the other hand is concerned with repeated bargaining. In his model, reference points do not only adjust upward but can be adjusted downward as well, i.e. $\forall\, a \in \mathbb{R}_+$: $f^+(a) = a$ and $f^-(a) = \beta\, a$ with $\beta \in [0, 1]$.

Shalev (2002) integrates reference points and loss aversion in cooperative bargaining games. With an extension of the Nash solution, the Kalai–Smorodinsky solution, and the Rubinstein game, he is able

---

[11] For the single-issue case, issue-specific indices are dropped.

[12] The notation differs from the original notation used by Compte and Jehiel to gain consistency with the attachment effect model.

to show that in his setup increasing loss aversion of one player leads to worse outcomes for that player. While the model by Shalev cannot easily be put in form of the update functions, the assumed origin of reference points is comparable: He suggests that reference points reflect aspirations and expectations that are formed from previous experiences and from knowledge of outcomes reached by others in similar situations. Furthermore, reference points might be influenced by the appearance, attitude, and behavior of the counterparty.

Linear Updates

A special case of the general attachment effect model is to assume linear updates of reference points as Compte and Jehiel (2006) suggest at the end of their paper and Hyndman (2005) formalizes it for downward adjustments of reference points. Formally, this can be written as $\forall \, a \in \mathbb{R}_{\mid} : f_k^+(a) = \beta_{1,k} \, a$, $f_k^-(a) = \beta_{2,k} \, a$, $g_k^+(a) = \beta_{3,k} \, a$, and $g_k^-(a) = \beta_{4,k} \, a$ with $\beta_{1,k}, \beta_{2,k}, \beta_{3,k}, \beta_{4,k} \in [0, \, 1]$ for all $k \in \{1, 2, \cdots, K\}$.

This case is especially interesting, as it includes the rational choice model as extreme case ($\forall \, k : \beta_{1,k} = \beta_{2,k} = \beta_{3,k} = \beta_{4,k} = 0$) and uses a minimum of parameters to allow (1) for all functions to vary depending on the difference of offer and reference point and (2) for the effect being differently pronounced depending on the party that makes the offer and an upward or downward adjustment. In Section 5.4, the parameters of such a linear model will be estimated based on experimental data.

3.3.4 Implications for Negotiations

If it is the case that

1. (some) negotiators evaluate offers and outcomes in a multi-issue negotiation relative to reference points,
2. reference points change depending on offers received as outlined above, and
3. the negotiators are loss-averse on single issues as outlined above,

one can predict the likely change in the negotiators' preferences during a negotiation. According to the attachment effect, the reference point moves towards the offers made during the negotiation on each issue. Thus, differences in an issue on which, for example, the counterparty frequently offers high values tend to be evaluated as losses rather than gains. Hence, such issues become more important in trading off different issues during the negotiation and in subsequent choices. The change of trade-offs can result in two polar cases: gains from trade that exist initially can be destroyed by negotiating or they can be created if they did not exist.

Destroying Gains from Trade

Figure 3.5 exemplifies the detrimental effect offers can have on mutual gains from finding a compromise agreement. Figure 3.5a shows the initial setup before a negotiation between S and H on issues A and B. Each party has a reference point and each party has an outside option, i.e. the BATNA, she will obtain when the negotiation fails to find an agreement. The figure sketches the indifference curves on which the respective BATNA's lie. S would prefer any outcome to the upper right from her indifference curve and H to the lower left of his indifference curve. Thus, the shaded 'lens' in the figure shows the set of mutually beneficial agreements that negotiators should try to find as both would be better off than with their outside options.

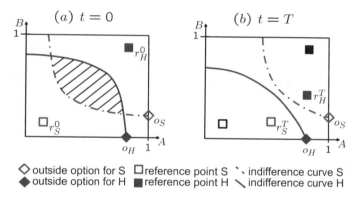

Fig. 3.5. Example of gains from trade being destroyed by the attachment effect

As there is incomplete information, the negotiators do not exactly know the set of mutually acceptable agreements. Furthermore, their preferences over these agreements are not identical and each party would like to claim as much of the joint gains as possible. Thus, the parties exchange offers and jointly explore the agreement space: they negotiate.

Both parties might make offers relatively close to their outside options, i.e. with high values on issue A and low values on issue B. This creates attachment to such agreements and the reference points gradually move to the lower right corner of the Edgeworth box; the new reference points r_S^T and r_H^T are displayed in Figure 3.5b. The shift of the reference points implies a change in the evaluation of some agreements as gains and losses; hence a change in the marginal rates of substitution, and, thus, new indifference curves. Negotiator S still prefers any agreement above her indifference that goes through her BATNA. This curve is, however, steeper now than it was at the beginning as—given the reference point r_S^T instead of r_S^0—differences on issue A tend to be evaluated as losses rather than gains. Hence the importance of issue A is increased for S and the indifference curve is steeper. The same reasoning applies for negotiator H: his reference point shifted from r_H^0 to r_H^T. As an effect, differences on issue B are perceived more severely by H. Overall, the shift of both parties' reference points destroyed any potential for a mutually beneficial agreement.

Creating Gains from Trade

The attachment effect can have the opposite implication as well: gains from trade can newly emerge during a negotiation. To exemplify this, Figure 3.6a sketches a situation in which initially no gains from trade exist between S and H. There is no agreement that would make either party better of than the respective BATNA.

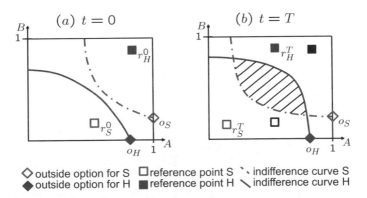

Fig. 3.6. Example of gains from trade being created by the attachment effect

Negotiating might alter this situation as the attachment effect might influence reference points. Figure 3.6b shows a situation in which H's reference point has shifted in a way that H now expects to obtain more on issue A for himself. Thus, an agreement giving H a low fraction of A is perceived as loss and this increases the importance of issue A for such outcomes. Hence, the marginal rate of substitution is changed. S, on the other hand, started with a relatively high reference point on issue A and lowered it in the course of the negotiation. Consequently, she tends to evaluate differences on A as gains and, thus, less strongly. Overall, the shift of reference points created a set of agreements that both parties prefer to their outside options. Thus, it is now mutually beneficial to find an agreement.

3.3.5 Summary

Negotiation analysis is an asymmetrically prescriptive/descriptive study of negotiations. Unlike game theory, it does not assume rationality of all parties but searches for descriptions of common patterns in negotiator behavior. Based on these, the prescriptive part suggests 'rational' non-equilibrium behavior given the likely cognition and actions of the counterparty. The descriptive part of negotiation analysis comprises a set of common biases in negotiations like the fixed pie illusion or escalation of conflict. The present study adds one such bias, the *attachment effect*.

In the process of negotiating, agents constantly perceive intermediate outcomes (offers), evaluate these, integrate them in their memory, and form expectations in the future course of the negotiation. According to the attachment effect, these expectations influence the agents' reference points, and hence, preferences. The strength of the attachment effect likely differs between individuals and issues. The attachment effect is not in conflict with the established behavioral patterns and biases in negotiations but introduces a new perspective on negotiator decision-making. All of these effects might interact with each other.

The functional relationship of offers and reference points presented above is one way of modeling an attachment effect in negotiations. It is, however, not the only one. Furthermore, literature suggests that an attachment effect might exist in negotiations but it has to be assessed whether this is really accurate. Ultimately, the success of the model will be assessed by empirical tests.

4

Internet Experiment

Experimental economists are leaving the reservation. They are recruiting subjects in the field rather than in the classroom, using field goods rather than induced valuations, and using field context rather than abstract terminology in instructions.

(Harrison and List, 2004, p. 1013)

Whether the attachment effect is present and reference-dependent preferences change systematically in multi-issue negotiations is empirically tested in two closely related experiments. In this chapter, an internet experiment that tests the existence of an attachment effect in negotiation is outlined. The results support the assumption that the preferences of negotiators are systematically affected by the offers exchanged. However—as with any experiment—the external validity of results might be questioned: an single experiment can never proof that the same results would emerge if any of the numerous design choices would be altered. To increase validity, a second experiment was conducted; it is reported in Chapter 5. The design of this second experiment is refined by lessons learned from the first experiment and purposefully differs in several respects to show that the attachment effect is not closely related to the design choices made.

The belief that experimentation is a proper methodology in economic research has grown dramatically over roughly the last 60 years. Nowadays, experimental economics is a widely accepted method of controlled data generation: see Guala (2005) for a recent methodological discussion, Davis and Holt (1992) and Friedman and Sunder (1994) for introductions on how to conduct experiments, and Kagel and Roth

(1995) for a handbook reviewing the most important fields tackled by experimental economists. See Reips (2002a,b) for an introduction to internet-based experimentation.

Harrison and List (2004) propose, upon observing a growing body of field experiment in the last years, that experimenters should be wary of the traditional wisdom that abstract, imposed treatments allow general inferences. In an attempt to ensure generality and control, researchers remove all field referents from the instructions and procedures. As a result—according to Harrison and List—the traditional lab experimenter has lost control to the extent that subjects seek to provide their own field referents. The present internet experiment 'leaves the reservation of experimental economics' with respect to two of the three factors addressed in the initial citation: a field good is used in a field context.

The chapter is structured as follows: Section 4.1 outlines the experiment design, discusses some of it's features, derives hypothesis on treatment effects based on the theory presented in previous chapters, and introduces the foundations of the statistical test to be used in the data analysis. Section 4.2 then presents the results; these favor the existence of an attachment effect in the present experiment. Section 4.3, finally, critically reviews the experiment design and points out changes made for the design of the lab experiment that will be described subsequently.

4.1 Experimental Design[1]

The experiment controls the course of a alternating offer multi-issue negotiation and uses a between-subject comparison of ex-post preferences to test for the existence of an attachment effect. Control over the negotiation is achieved by determining the strategy of a software agent that takes the role of one negotiator and exchanges offers with a subject. The object of negotiation is a hypothetical tenancy contract, i.e. students negotiate over the terms of a contract for renting an apartment. The contract is, however, hypothetical and not really signed at the end.

A major challenge in testing for a change of preferences is, that negotiators' preferences are neither directly observable, nor can they be elicited reliably (at least not multiple times from the same subject). Thus, a within-subject comparison of preferences at different times is

---

[1] Parts of Sections 4.1 and 4.2 are closely related to Gimpel (2007).

not possible and a between-subject comparison is applied.[2] A between-subject comparison has difficulties as well, as preferences naturally differ across subjects. However, with randomized assignment to treatments and with a large enough sample these individual differences are expected to equal out.

The main idea of the experiment is to let subjects in different treatments face counterparties with different negotiation strategies and to measure their preferences after the negotiation. As the assignment of subjects to treatments is randomized, there should be no systematic differences in the subjects' ex-post preferences between treatments except if the negotiation process itself influences preferences. Thus, systematic differences between treatments favor the hypothesis that preferences are affected by the attachment effect.

4.1.1 Procedure

Subjects were recruited in an undergraduate class on business administration: the lecturer promoted participation and sheets of paper with an account name and password were handed out. Students were informed that they would need about 10 to 15 minutes for participating. They then logged in from home over the internet via a web browser. As it is standard for internet experiments, information was presented via the subject's computer screen, subjects' actions were recorded at the web server in log files, and log files were formatted and filtered to the requirements of statistical analysis (Reips, 2002b). Loading times of web pages are important, as a slow process might cause drop outs. All pages could be loaded in less than a second with a standard modem connection to the internet.

Incentives

A week after recruiting the subjects, 50 euros were awarded to one of the participants in a lottery held in the lecture. Each student who completed the experiment had the same chance of winning the lottery which was not related to the subjects' specific choices or 'success' during the experiment. Therefore, the lottery served as an incentive for

---

[2] Curhan, Neale, and Ross (2004) use a within-subject comparison of preferences in an experiment on endogenously changing preferences in negotiations. Preference elicitation for each of ten potential contracts is achieved by subjects rating these contracts on Likert scales. To reduce consistency induced by subjects awareness of multiple preference elicitations, the order of the contracts to be rated varied from elicitation to elicitation.

participation—it was not a salient reward as oftentimes used in experimental economics (Smith, 1982a; Davis and Holt, 1992, Ch. 1).

An important point in analyzing the data gathered in the experiment is whether subjects were sufficiently motivated not only for participation per se but to sincerely consider the choices presented to them. This might be questioned, as the financial reward was non-salient. Casual observations indicate that subjects were motivated. Firstly, immediately accepting the counterparty's first offer would have been the fastest way to secure participation in the lottery. However, none of the subjects chose this least-effort-way and just one subject accepted the agent's second offer. All other subjects negotiated eagerly.

Secondly, overall 47 students logged in on the experiment's web site; these are about 90% of the students addressed in the lecture. In four cases the session was abandoned by the subject and one observation had to be discarded as the subject used the forward and backward functionality of its web browser.[3] It is not analyzed in the following. In the instructions, subjects were explicitly told not to use these functions as this could allow retracting to an offer by the agent for accepting it after already seeing the agent's next offer. Overall, there are 42 valid observations and it appears that the subjects' intrinsic motivation for participating sincerely was rather good.

It is highly debated whether salient rewards are necessary in experimental economics; see e.g. Gneezy and Rustichini (2000), Read (2005) and Guala (2005, Ch. 11) for discussions of this methodological question. The basic assumption here is that the reward structure—which is identical for all subjects—does not induce a systematic treatment difference. Furthermore, the topic is not pivotal, as the corresponding lab experiment uses salient rewards and corroborates the results (cf. Ch. 5).

Control over the Subject Pool

One of the benefits of internet experiments is oftentimes seen in the increasing number of users to which the experimenter has relatively easy access. Self selection of internet users is, however, a serious problem. Neither the advantage nor the disadvantage are important for the present experiment as subjects were recruited in the classroom, every student approached in the lecture has internet access (either at home or at the university), and about 90% of the students participated.

With internet experiments, a danger lies in people participating multiple times (Smith and Leigh, 1997). Three precautions were taken to

---

[3] The discarded data is reported in an appendix which is available upon request.

avoid this. The individual accounts handed out in the lecture are one, as a single account did not allow participating multiple times. Secondly, subjects had to enter their name and an e-mail address to participate in the lottery. Thirdly, the IP addresses of the subjects' computers were recorded to dismiss observations if multiple log-ins would occur from the same address (Reips, 2002b). No name, e-mail address, or IP address occurred twice. Furthermore, previous research has shown that people invited for participation only very rarely disregard the request to take part only once (Reips, 2000). Thus, it is assumed that no subject participated twice and that observations are independent of one another.

Matching Procedures

Each subject negotiated with a software agent. Participation was independent of the other subjects, i.e. the decisions made by one subject did not influence the other subjects.

Timing

The experiment took place from April 27 to May 3, 2005, at the University of Karlsruhe. A typical session lasted about 10 to 15 minutes. If a user would have been inactive for more than 20 minutes during her session, the software would have reported a failure. This did not occur.

4.1.2 Course of a Session

A subject's session starts with two pages of instructions followed by a short questionnaire to ensure understanding. Then the subject bilaterally negotiates with a software agent over a hypothetical tenancy contract. The subjects' instructions explain the entire procedure. Subjects are told that they negotiate with software agents.[4]

Object of Negotiation

Subjects negotiate about a tenancy contract: The monthly rent in euros (denoted as R), the availability of an elevator (A), and the existence of a balcony (B) are the three decision criteria that are up to negotiation. All other attributes like the size of the apartment, the available furniture, the location, etc. are non-negotiable and fixed in the subjects'

---

[4] The full instructions, the questionnaire to check for understanding, and screen shots from the system are given in an appendix which is available upon request.

instructions. The terms 'issues' and 'attributes' are used interchangeably for the rest of this chapter, as negotiation issues are attributes of the tenancy contract in this experiment.

The agreement space in the experiment is $X = \{x = \langle R, A, B \rangle \mid R \in \{147, 161, 176, 193, 215, 232, 259, 274, 298, 323\}, \ A, B \in \{yes, \ no\}\}$. All rent contracts are hypothetical, i.e. no subject signs a real binding contract. Thus, there is no salient reward in this experiment.

The tenancy contract is chosen as object of negotiation for two reasons: Firstly, it is expected that the students can relate to this object as most of them had to look for an apartment when they recently moved to Karlsruhe. Thus, they are able to transfer their preferences from the field. Secondly, it is expected that subjects do not have to much experience with this domain and preferences are not yet stabilized (in the terminology of constructive consumer choice; cf. Sec. 2.3.1). Most students did not negotiate tenancy contracts often enough to have a precise idea on the value of an elevator or a balcony. Thus, there might be potential to influence preferences.

Negotiation Protocol

The negotiation protocol is an alternating offer multi-issue negotiation. During the negotiation, offers are exchanged between the two parties. A software agent representing the landlord starts with an initial offer and subsequently the parties alternate in deciding whether they accept their counterparty's offer or they prefer to propose a counteroffer. Offers are points in the agreement space and they are exchanged electronically. No argumentation or other free text, audio, or visual signals are possible. Figure 4.1 shows a screen shot from the alternating offer negotiation.[5] Students were used to the general layout, i.e. the images at the top and the color scheme, as it resembles the web site of the lecture in which they were recruited. The screen comprises three sections: in the upper part, the scenario is explained; in the middle part the subject sees the landlord's current offer and can either accept it or enter a counteroffer; at the lower end, all previous offers are displayed. In the lower left corner there is a link to the experiment's instructions.

If either party accepts an offer by the counterparty, this offer becomes the agreement between the subject and the first landlord. If a maximum of overall twelve offers is reached—i.e. six offers per party—without any of the offers being accepted, a third party steps in and proposes an agreement which is binding for the two parties. This third

---

[5] The text in this and the following screen shots is translated. The original texts are given in an appendix which is available upon request.

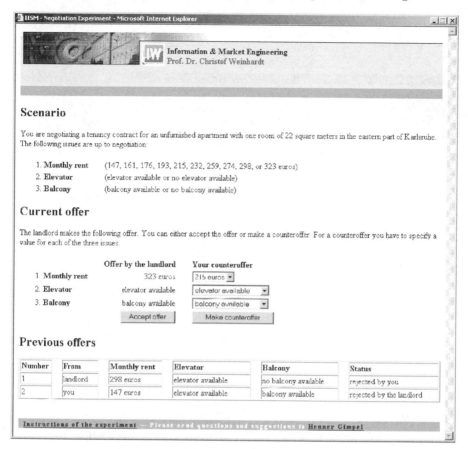

Fig. 4.1. Alternating offer negotiation (screen shot)

party is called *arbitrator* in the following. An arbitrator generally has the authority to impose a solution. This differentiates him from a mediator, for example, who can help with the negotiation process but cannot dictate an agreement (Raiffa, 1982, Ch. 2).

Wealth Effects

After the negotiation with the first landlord ends, a second landlord makes a single offer for an alternative contract. The subject can either reject the second landlord's offer and stick with the agreement she has with the first landlord, or she can accept the second offer. In the latter case, the agreement with the first landlord is automatically resolved. No matter whether the subject chooses the agreement with the first or the offer from the second landlord, the result is called *final contract.*

Figure 4.2 shows a typical screen on which the subject can choose her final contract.

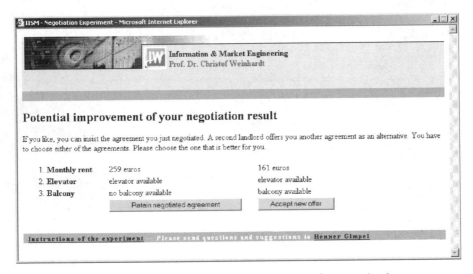

Fig. 4.2. Offer by the second landlord (screen shot)

Under the assumption that subjects prefer a low monthly rent, having an elevator, and having a balcony, the offer by the second landlord is by design strictly better for the subject than any agreement she could have with the first landlord (without the subject knowing this in advance). Thus, it is expected that all subjects accept the second landlord's offer. It serves for giving pairs of subjects in different treatments the same final contract and, thus, to level out all objective differences between subjects that might arise from the alternating offer negotiation. Individual reference points might, however, remain. This allows for a blocked analysis of the data. Indeed, each single subject accepted the second landlord's offer.

Measurement of Ex-Post Preferences

After the interaction with the two landlords, the subjects' willingness to accept a change in a single attribute of the final contract is elicited and compared between treatments. Given the final contract, a subject's willingness to accept a worsening in either the attribute elevator (A) or the attribute balcony (B) is elicited by asking the subject to solve two indifference equations for the respective monthly rent, i.e. the monthly rent is taken as numeraire. With respect to the response modes discussed in Sections 2.1.2 and 2.3.2, the indifference equations are *matching mode*. The respective computer screen is shown in Figure 4.3.

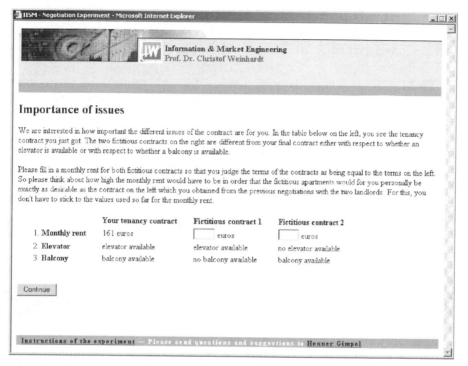

Fig. 4.3. Preference elicitation (screen shot)

The indifference equations to solve are

$$\langle R_i, yes, yes \rangle \sim \langle R_{i,B}, yes, no \rangle \sim \langle R_{i,A}, no, yes \rangle$$

R_i is given via the offer of the second landlord and the subject enters $R_{i,A}$ and $R_{i,B} \in [0, 999]$. The indifference equations elicit the subjects' ex-post preferences,i.e. preferences after having negotiated with agent A. As the values of $R_{i,A}$ and $R_{i,B}$ likely depend on R_i, transforming them increases comparability. To this end, a subject's willingness to accept (*WTA*) a worsening in either attribute A or B is calculated. They are defined as follows:

$$WTA_{i,A} = R_i - R_{i,A} \quad \text{and} \quad WTA_{i,B} = R_i - R_{i,B}$$

These *WTA* values are analyzed to test for a treatment effect. The rational choice model along with random assignment of subjects to treatments suggests that the *WTA* values should not differ systematically between treatments. The attachment effect model, on the contrary, predicts a systematic dependence of preferences on offers, and hence, a dependence of *WTA* values on treatments. This will be discussed in more detail in Section 4.1.5. Before that, however, the strategies of the first and the second landlord are outlined and some design features are reviewed critically to analyze more closely how the design might influence results.

Finally, a session ends with asking for the name and the e-mail address, thanking the subject, and asking for general comments on the experiment.

4.1.3 Treatments and Agent Strategies

The strategy of the software agent taking the role of the first landlord is to present a fixed offer sequence one offer at a time until either (1) the subject accepts an offer; (2) the agent accepts an offer by the subject; (3) overall twelve offers are exchanged. The agent accepts a subject's offer if it weakly dominates either one of the agent's previous offers or the agent's next offer. Weak dominance means that the subject asks for no more than the agent offers, i.e. she does not ask for an elevator that the agent does not offer, not for a balcony that the agent does not offer, and not for a lower rent than the agent offers.

The experiment embraces two treatments (*T1* and *T2*), each treatment has two offer sequences. The offer sequences are determined prior to the experiment. In the first sequences used in treatment 1 (*T1'*), each single offer grants an elevator to the subject, i.e. $A = yes$. The values of the other attributes are randomized with a slight tendency to making concessions as the negotiation progresses. The offer sequence is displayed in Table 4.1 under the heading *T1'*. For *T1'*, attribute A is termed *reference attribute* and attribute B is termed *non-reference attribute*—this labeling is indicated in the table's last line.

The second offer sequence used in treatment 1 (*T1"*) is derived from *T1'* by exchanging the values of issues A and B. Thus, $B = yes$ for each of the agent's offers and the other two attributes take the randomized values from the first offer sequence. The offer sequences in *T2* are constructed as follows: for *T2'*, no offer grants an elevator to the subject and values for the other two issues are taken from *T1'* and, analogously, *T2"* is derived from *T1"* by setting attribute B to *no* for all offers. All offer sequences are given in Table 4.1. The subjects' individual responses are outlined in an appendix which is available upon request. Both offer sequences in treatment 1 have in common

Table 4.1. Offer sequences used by the software agent

| Offer number | treatment 1 | | | | | | treatment 2 | | | | | |
| | $T1'$ | | | $T1''$ | | | $T2'$ | | | $T2''$ | | |
	R	A	B	R	A	B	R	A	B	R	A	B
1	298	*yes*	*no*	298	*no*	*yes*	298	*no*	*no*	298	*no*	*no*
2	323	*yes*	*yes*	323	*yes*	*yes*	323	*no*	*yes*	323	*yes*	*no*
3	298	*yes*	*yes*	298	*yes*	*yes*	298	*no*	*yes*	298	*yes*	*no*
4	259	*yes*	*no*	259	*no*	*yes*	259	*no*	*no*	259	*no*	*no*
5	259	*yes*	*no*	259	*no*	*yes*	259	*no*	*no*	259	*no*	*no*
6	274	*yes*	*yes*	274	*yes*	*yes*	274	*no*	*yes*	274	*yes*	*no*
Ref.	–	*yes*	*no*	–	*no*	*yes*	–	*yes*	*no*	–	*no*	*yes*

that one attribute has the value *yes* for each offer. The offer sequences in treatment 2 have in common, that the respective attribute has the value *no* for each offer.

The second landlord offers $\langle R_i, yes, yes \rangle$, where R_i is randomized across pairs of subjects and the subscript denotes a single subject. As this offer was accepted by all subjects, it is the final contract for all subjects.

4.1.4 Discussion of Design Features

Internet experiments generally allow less control over the conditions under which participants complete the experiment than traditional lab experiment do. The noise, distraction, mood, fatigue, motivation, etc. of subjects cannot be assessed.[6] According to Musch and Reips (2000)—who discuss internet experiments in psychology—these extraneous factors are the biggest concerns of experimenters as they increase the variability of data and jeopardize internal validity. This is, however, as well a virtue of non-laboratory experiments, as it increases external validity, i.e. transferability, as Musch and Klauer (2002) and Harrison and List (2004) point out.

The absence of any salient reward in the present experiment was already addressed above. Like the relaxed control of an internet experiment, it is not of much concern here as the lab experiment reported in the next chapter differs with respect to these two design decisions and affirms the results obtained here.

---

[6] Note that several of these factors, like mood for example, cannot be controlled in the lab as well.

Software Agents

The use of software agents as counterparties in the negotiations instead of having negotiations between two human subjects deserves are more thorough discussion. The benefits of this design decision are:

1. The experimenter has absolute control over the offers made by one party and, thus, over the differentiation between treatments.
2. The design does not require coordinating the timing of subjects' participation or long delays in the offer exchange. Instead, the agent can immediately reply to a subject's offer.
3. Each single subject contributes an independent observation to the data without any confounding influence from the interaction with other subjects. This increases the sample size for statistical analysis compared to solely recording interaction among subjects.[7]

The first argument is pivotal, the second a nice side effect, and the third argument just reduces costs of conducting the experiment (or, given a fixed prize in the lottery, it increases the subjects' expected payoff and thus the incentive for participation).

On the other hand, the usage of software agents has a drawback: Would subjects behave similarly if they would negotiate with human counterparts? As long as this question cannot be answered, it cannot be claimed that the experiment's results generalize to a wider spectrum of situations, namely negotiations among humans.

Alternative Designs

The major alternative would be to have pairs of subjects negotiating with each other: On the one hand, both negotiators could be free to choose any offer they like (as subjects are in the present design). This would, however, mean to sacrifice the distinction between treatments and to solely apply an ex-post measure to classify negotiations. On the other hand, one of the negotiators could be restricted in the offers she makes. As an extreme case, she could be required—as the agent is—to follow the exact same strategy that is predetermined by the experimenter. This design is chosen by, for example, Sanfey et al. (2003) to compare how subjects play an ultimatum game against a human or

---

[7] Turel (2006) notes, for example, that the fact that most negotiation data is collected from dyads of negotiators has statistical implications for the analysis. Simply assuming independency of individuals—as it is done by some researchers—can lead to incorrect inferences. See as well Kenny and Judd (1986, 1996) for a discussion of the statistical challenge.

a computer taking the exact same actions: 'Offers made by human partners in fact adhered to a predetermined algorithm, which ensured that all participants saw the same set (and a full range) of offers [...]' (p. 1756). However, this design was not chosen, as this could be interpreted as deceiving the subjects which is generally proscribed in experimental economics. See e.g. Friedman and Sunder (1994, Ch. 4.7) and Hertwig and Ortmann (2001) for a discussion of this ethical issue.[8]

Another alternative design would be to let human subjects play with computer opponents without informing them about the nature of their opponent. This design is chosen by, for example, Roth and Schoumaker (1983) and Rilling et al. (2002). As this concealment could be interpreted as deception as well, subjects were accurately informed that they would interact with software agents.

Playing against Computers

Computers have already been used in the 1960s and 70s as players in experiments on zero-sum games (Lieberman, 1962; Messick, 1967; Fox, 1972). Insights on how the computer opponent influences the subjects' strategies can be gained from experiments in which subjects face human as well as automated counterparts. These experiments differ with respect to whether or not the subjects know the computer's (probabilistic) strategy for sure.

Walker, Smith, and Cox (1987) employ automated bidders in a first price sealed bid auction. Subjects know that they are playing computers. The computer's strategy simply is—without the subjects knowing this—to bid the Nash equilibrium strategy. No significant difference in the behavior of subjects depending on the nature of the opponent can be shown. Walker, Smith, and Cox (1987, p. 244) present their approach as 'initial investigation into the role of computerized competitors as a methodological tool for testing Nash bidding theories' and see that their results support 'the applicability of this methodology to alternative market environments.' Glimcher, Dorris, and Bayer (2005) let their subjects play several hundred repetitions of an inspection game amongst an employer and an employee. Participants play against either another human or a computer. The authors find that the computer elicited behavior from the human counterpart that is statis-

---

[8] Sanfey et al. (2003, p. 1758) address the issue of deception in their endnotes 14 and 15. They classify this procedure as 'limited amount of deception' that they purposefully included in their design primarily to reduce costs and logistic demands of conducting the experiment.

tically indistinguishable from the behavior when playing against other humans.[9]

Fehr and Tyran (2005) not only informed their subjects about the nature of their opponent, but also the strategy the computer opponent would play. Thus, in this experiment on money illusion, there was no strategic uncertainty. The subjects' behavior differed between opponents, i.e. the level of strategic uncertainty, and the efficient equilibrium is reached more often against computers playing a publicly known strategy. Similarly, Bearden, Schulz-Mahlendorf, and Huettel (2006) let a subset of their subjects play a simplified poker game against a computer opponent whose strategy is known. Again, subjects face a straightforward optimization problem without any strategic uncertainty. Not surprisingly, the authors report that the computer's fixed and known policy changes subject behavior.

Neuroeconomics

Sanfey et al. (2003) made a within-subject comparison how subjects play the responder role in an ultimatum game against human opponents or computers.[10] The behavioral results, i.e. the subjects' acceptance or rejection of offers, for playing against humans is in line with the finding from other experiments: responders accept all fair offers and show a decreasing acceptance rate as the offers become less fair to their disadvantage. Unfair offers made by subjects are rejected significantly more often than the same unfair offers made by the computer. In this experiment, the rejection of an unfair offer from a human cannot induce a more equal split in a subsequent round, as each subject played with each human proposer just once. Even more puzzling is the rejection of offers by the computer except if the subjects tried to convey their unhappiness to the experimenter.

Furthermore, using fMRI, Sanfey et al. identify brain regions that are activated in the evaluation of unfair offers (bilateral anterior insula, dorsolateral prefrontal cortex, and anterior cingulate cortex) and report that the magnitude of activation of these regions was significantly greater for unfair offers from human proposers than for the same unfair offers from the computer. Sanfey et al. conclude that players not only react to the amount offered, but are sensitive to the context, i.e. they

---

[9] The sample size in this study by Glimcher et al. is, however, quite small like in many neuro-imaging experiments: just 8 human subjects playing another human and 8 subjects playing the computer.

[10] This is the aforementioned study in which some human players adhered to a predetermined algorithm.

have a stronger emotional reaction to unfair offers from humans than to the same offers from a computer.

Several authors report that the endowment effect and loss aversion are more pronounced in emotion-laden situations (Horowitz and Mc-Connell, 2002; Lerner, Small, and Loewenstein, 2004; Loewenstein and O'Donoghue, 2005). The same might be expected for the attachment effect. If so, finding an attachment effect in negotiations against software agents that are likely not perceived very emotionally by subjects might be viewed as especially strong result.

In a study on female subjects playing a series of prisoner's dilemma games, Rilling et al. (2002) employ computers that initially defect and then play tit-for-tat (without the subjects knowing this strategy). Some of the subjects were told that their opponent would be a computer, others were not. For subjects that knew about the true nature of their counterparty, mutual cooperation was less common throughout the game (although the computer's strategy was exactly the same in both settings). Furthermore, the authors used fMRI neuro-imaging and conclude—as Sanfey et al.—that playing a computer activates partially different brain regions than playing another human.

In another neuro-imaging study on a two-person trust and reci-procity game, McCabe et al. (2001) report differences in activation of brain regions for cooperators depending on whether they play with a human or a computer opponent (with a fixed and known probabilistic strategy). Within the group of cooperators, i.e. subjects that relatively often cooperate with their counterpart in the trust game, regions of the prefrontal cortex are more active when subjects are playing a hu-man opponent than when they are playing against a computer. For non-cooperators, the authors did not find any difference.[11]

Further economic experiments in which subjects played against com-puters are reported by Kiesler, Sproull, and Waters (1996), Houser and Kurzban (2002), Shachat and Swarthout (2002), Shachat and Swarthout (2004), Winter and Zamir (2005), Duersch et al. (2005), Kirchkamp and Nagel (2007).

To summarize, the usage of computer opponents in experimental economics is not uncommon. Several studies have approached the ques-tion on how behavior differs depending on whether one or several other players are computers or humans. In some studies it was found that (1) playing a computer is different when all strategic uncertainty is removed, (2) building up social relationships like trust and coopera-

---

[11] Again, the subject pool is relatively small: 7 out of 12 subjects were classified as cooperators and the remaining 5 as non-cooperators.

tion is more common with other humans than with computers, and (3) that actions of a human opponent are evaluated more emotionally than actions of a computer and, thus, the attachment effect might be less likely to appear with computer agents. Nevertheless, all cited works study isolated effects and, to date, there is no general model on how behavior depends on the nature of the counterparty although it is easily imaginable that the interaction with computers in electronic markets will increase in the future.

Strategic Considerations

What does it mean to negotiate with a software agent? Does a software agent possess preferences? And which strategy will it use? A computer itself obviously does not have preferences; however, it likely represents the preferences of its programmer (given the assumption that the computer was programmed properly). Thus, interacting with the computer is like interacting with a (human) agent to whom a principal delegated the task of negotiating with an explicit preference ordering and a given (maybe probabilistic) strategy. This delegation is like the strategy method employed in many experiments (Selten, 1967): a subject is required to specify her entire strategy, i.e. a concise plan of actions for all possible situations, and this strategy is then played by a proxy to determine the outcome.

So the question about the agent's strategy becomes a question concerning the principal's strategy. Social and psychological aspects aside, for strategic considerations, negotiating with a software agent is not much different from negotiating with an employee of a landlord or a car dealer, for example. The employee of a car dealer himself does not have preferences for the different aspects of the sales contract. However, the customer assumes (or knows) that the owner of the shop specified his preferences, e.g. via a reserve price and the employee's premium, and gives out a policy how to negotiate and which additional features to offer in which situation. Whether the human agent's actions depend on an incentive scheme or the software agent's behavior on a fixed program might not matter much for the counterparty in a negotiation.

Realism

Most real life negotiations certainly take place between humans. Thus, negotiating with a software agent might be perceived as odd. However, the interaction with computers becomes more and more an ordinary activity. Playing games against computers is a popular hobby and in chess, for example, the best computer programs outperform even human world champions (Campbell, Hoane, and Hsu, 2002). Furthermore, the interaction of humans and (more or less sophisticated) computer counterparts in markets is not uncommon.

Internet auction sites like eBay.com and Amazon.com employ proxy bidding: the bidder specifies the maximum bid she is willing to submit and the software then takes her place in an English auction. Each time the agent is outbid and as long as the upper limit is not exceeded, the agent submits a new bid just a minimum increment above the standing price. Thus, bidders in these auctions do not bid directly against other humans, but against software agents. The same is true for so called sniping agents submitting bids to eBay auctions in the last minutes or seconds of an auction.

On other internet sites, like Priceline.com, humans bargain with software agents. A customer intending to purchase a flight, for example, submits an offer to the site, i.e. she specifies a schedule and a price. The offer is either accepted or rejected by a software agent—Priceline.com terms this mechanism 'Name Your Own Price'. If it is accepted, the customer has to pay the price she offered, if it is rejected she has to wait for at least seven days before she can submit a new offer for the same flight schedule. Finally, software agents and algorithmic trading automata are used by many institutional investors in financial markets.

Summary

The usage of software agents increases the experimenter's control over the negotiations—it increases internal validity. On the other hand, it limits external validity. It is not clear how subject's perceive the play against a computer rather than a human and, even if the perception would be obvious, it is not clear how the subjects' strategic considerations, emotions, and actions depend on the nature of the counterparty. However, the interaction with software agents likely is not uncommon for many of the student subjects and the usage of computer counterparts likely reduces the effect of other regarding preferences and emotions. If the finding that loss aversion is more pronounced in emotion-laden situation carries over to the attachment effect, this would suggest that finding an attachment effect with the present design is an especially strong result.

4.1.5 Hypothesis on Treatment Effects

The attachment effect model presented in Section 3.3 for the emergence of issue-wise reference points in a negotiation is based on issue-specific and subject-specific update functions. For illustrating the following hypothesis, it is drastically simplified: assign the value zero to *no* on issues A and B and the value unity to *yes*. Furthermore, assume that $f_k^+(\cdot)$ $k \in \{A, B\}$ are monotonically increasing functions jointly applying for all subjects and that all other update functions and the initial reference point are equivalent to zero for issues A and B ($f_k^-(\cdot) = 0$, $g_k^+(\cdot) = 0$, $g_k^-(\cdot) = 0$, $r_k^0 = 0 \; \forall \; k \in \{A, B\}$ for each subject). No special assumption concerning issue R is needed here. This simplification is a sufficient assumption for the following hypotheses, its restrictiveness is, however, not necessary for deriving the predictions. The model is solely used for illustration; necessary assumptions are pointed out below.

Given the above specification of update functions and the offer sequences used in the experiment, one can hypothesize about the reference point of every subject with respect to issues A and B. Obviously, the precise reference point depends on the form of $f_A^+(\cdot)$ and $f_B^+(\cdot)$, but the relative location of the hypothesized reference points for different subjects can be calculated without assuming specific update functions.

Hypothetical Reference Points

Figure 4.4 displays these relative locations with respect to attributes A and B. The axes do not have any marks in between 0 and 1, as this would depend on the functional form of the respective $f_k(\cdot)$.[12] The third attribute, i.e. the monthly rent, is omitted here, as its values do not differ between treatments. Solid circles represent hypothesized reference points stemming from offer sequences *T1'*, squares from *T1"*, triangles from *T2'*, and diamonds stand for hypothesized reference points arising from *T2"*. One clearly sees that *T1"* and *T2"* are derived from *T1'* and *T2"*, respectively, by exchanging issues A and B. (In the figure, they are mirrored at the diagonal from the lower left to the upper right corner.) For each single offer sequence, there are different possible reference points, as the duration of a negotiation is determined by the subject. In all treatments, the reference point starts at the origin and gradually moves towards the upper right corner when the subject receives an offer by the agent. Furthermore, the offer by the second landlord likely has a strong influence on the subjects' reference points and, in Figure 4.4 moves them to the upper right corner. However, it is omitted here, as the influence is the same in both treatments.

The numbers close to the potential reference points indicate how many subjects are expected to end at this specific point. This infor-

---

[12] In fact, the figure is drawn with respect to the functions $f_k^+(a) = \frac{1}{4} a$ for $a \geq 0$ and $k \in \{A, B\}$.

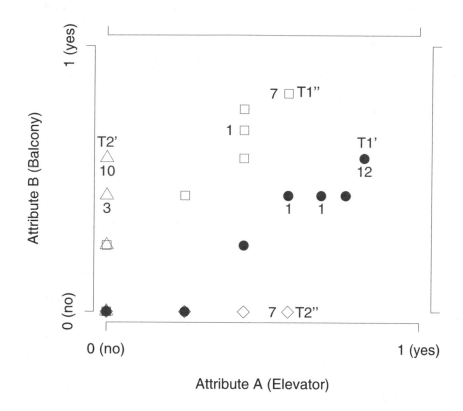

Fig. 4.4. Hypothesized reference points for different subjects given the offer sequences

mation takes the observed duration of each subject's negotiation in account.

Figure 4.4 suggests that the reference points of subjects playing against *T1'* and *T2'* are about the same with respect to attribute *B* but very different on attribute *A*. Attribute *A* is termed *reference attribute* for *T1'* and *T2'*, as noted before. With respect to attribute *B*, no difference between *T1'* and *T2'* would be expected from the figure. Attribute *B* is termed *non-reference attribute* for the two offer sequences.

Given the hypothetical reference points in the figure, subjects confronted with *T1'* are expected to feel more entitled to getting an apartment with an elevator than subjects facing *T2'*. The measure WTA_A captures how strongly subjects perceive a change from an apartment with balcony and elevator (the upper right corner of Figure 4.4) to an apartment with balcony but without elevator (the upper left corner of the figure). For subjects having played $T1'$, the loss of an elevator presumably looms larger than for subject having played $T2'$. Thus, given the reference point model, $WTA_A^{T1'} > WTA_A^{T2'}$ is expected.

For offer sequences $T1''$ and $T2''$ the labeling of reference and non-reference attributes works analogously by exchanging attributes A and B. Thus, observations from $T1'$ and $T1''$ can be grouped together as well as observations from $T2'$ and $T2''$. The differentiation between attributes is then concerning the reference attribute (A for $T1'$ and $T2'$, B for $T1''$ and $T2''$) and the non-reference attribute (B for $T1'$ and $T2'$, A for $T1''$ and $T2''$).

Predictions

Overall, the attachment effect model suggests the following predictions, as exemplified by Figure 4.4:

$$WTA_{Ref}^{T1} > WTA_{Ref}^{T2} \quad \text{and} \quad WTA_{NRef}^{T1} = WTA_{NRef}^{T2}$$

where the subscript *Ref* indicates the reference attribute and *NRef* the non-reference attribute.

On the contrary, a rational choice model without a change of reference points predicts that the willingness to accept should be independent of the treatment. More formally, this means that the two equations

$$WTA_{Ref}^{T1} = WTA_{Ref}^{T2} \quad \text{and} \quad WTA_{NRef}^{T1} = WTA_{NRef}^{T2}$$

should hold.

Necessary Assumptions

So far, the presentation was derived from a simplified version of the attachment effect model presented in Section 3.3. This is however, not necessary for the predictions. It is only necessary to assume that in the behavioral model the reference point tends to move towards the counterparty's offers.

Consider a rational choice model first: Either there is no reference point and, thus, no loss aversion, or the model might allow for subjects' reference points as long as they are static. Anyways, in rational choice models it is assumed that the offer sequences do not have any impact on the subjects' reference points. Together with random assignment of subjects to treatments this implies that reference points do not differ systematically between treatments (if they exist at all). Thus, given a single issue, subjects' willingness to accept should be the same (up to random disturbances) in all treatments. This is expressed in the equations given above.

The attachment effect model that allows for a change of reference points due to the alternating offer exchange suggests otherwise. Assume the subjects' reference points tend to gradually move towards the value offered by the agent and that there is no influence of the subjects' own offers that systematically differs between treatments. No matter what the initial reference points are, with random assignment to treatments, in *T1* the reference points will tend to be higher with respect to the reference issue than in *T2*. The reason is that the reference issue is offered by the agent in every single offer in *T1* whereas it is not offered in a single offer in *T2*. This prediction is expressed in the inequality given above. The non-reference issue and issue *R* are equally randomized for both treatments. Thus, no difference is expected here.

4.1.6 Foundations of the Analysis

The statistical test employed is a rank-based exact permutation test. Permutation tests were introduced in the 1930s by Fisher (1935a,b) and Pitman (1937, 1938); frequently used synonyms are randomization test and re-randomization test. Permutation tests were among the very first statistical tests to be developed—they were, however, beyond the computing capacities of the 1930's for all but very small sample sizes. Hence, over decades statisticians proofed asymptotic behavior of parametric tests, as they had not enough computing power to calculate the demanding exact permutations (Good, 2000, Ch. 1). Nowadays, computing power has increased that much that permutation tests are applicable to a wider range of data sets as, for example, the one collected in the present experiment. See Sheskin (2004, Test 12a) for a short and Good (2000, esp. Ch. 1–3, 9–11, and 14) for an extensive and excellent overview on permutation tests.

In a permutation test, the reference distribution for the test statistic is obtained by calculating all possible test statistics for the given data. This is done by permuting the observed data points across all possible outcomes, given a set of conditions consistent with the null hypothesis. With this, it is not necessary to rely on the existence of a hypothetical infinite population and to approximate the distribution of the test statistic by a common distribution like the normal or the χ^2 distribution. Instead, exact p-values can be computed.

4.2 Experimental Results

Table 4.2 displays the subjects' willingness to accept values which follow from the two indifference equations solved by every subject. Each line represents data from two subjects in the two different treatments.[13] Column 1 shows the offer sequence as well as the monthly rent R_i that a subject has in its final contract; column 2 gives the subjects' identification numbers. The data is grouped by offer sequences and R_i, as these factors likely introduce variability in the elicited WTA values. The pairwise display of two specific subjects in a single line is for layout reasons. Data from subject 13, for example, are in one line with subject 49's data; it could as well have been in the same line with data from either subject 63 or 64: the grouping by offer sequences and R_i is relevant, the lines are not.

Overall, there are 20 subjects in *T1* and 22 in *T2*. In *T1*, for example, there are three subjects (numbered 13, 69, and 98) that had a contract for an apartment with both an elevator and a balcony and a rent of 161 euros when they were asked to specify equally desirable rent contracts for apartments that either do not have an elevator (column 3) or no balcony (column 5). Subject 13, for example, chose a monthly rent of € 150 for an apartment without availability of an elevator but with balcony. Accordingly, for this subject WTA_A is 11; the value is given in column 3 of the table. Remember that in *T1'*, *A* is the reference attribute. Ranks of observations are given in parentheses besides the WTA values.

Non-Reference Attribute

The predictions of the rational choice model and the attachment effect model coincide for the non-reference attribute: there shouldn't be a treatment effect. Mean and median values of WTA—given at the lower end of columns 5 and 6—support this hypothesis as they are approximately equal. A two-sided rank-based exact permutation test is employed to test for significance. This test accounts for (1) the special grouping of observations by offer sequences and R_i, (2) outliers, (3) the non-normality of the data, (4) ties, and (5) the relatively small sample size that renders approximate parametric methods inappropriate.

The null hypothesis of no treatment effect for the non-reference attribute cannot be rejected (p-value > 0.5). Hence, for the non-reference attribute, no significant difference between treatments is found.

---

[13] There are two exceptions for which no observation in *T1* exists.

RESULT 4.1: *As predicted by both models, there is no treatment effect for the attribute on which the agent's offer sequences are randomized (the non-reference attribute).*

Table 4.2. *WTA* values elicited in the experiment

1	2	3		4		5		6	
Of. sequence (final R_i)	Subject numbers	WTA_{Ref}^{T1} (rank)		WTA_{Ref}^{T2} (rank)		WTA_{NRef}^{T1} (rank)		WTA_{NRef}^{T2} (rank)	
T1'/T2'	13 / 49	11	(6)	6	(3)	11	(3.5)	31	(6)
(161)	69 / 63	10	(5)	0	(1.5)	30	(5)	6	(1)
	98 / 64	8	(4)	0	(1.5)	11	(3.5)	10	(2)
T1'/T2'	53 / 74	8	(3)	6	(2)	10	(1)	16	(2.5)
(176)	/ 80			0	(1)			16	(2.5)
T1'/T2'	75 / 54	13	(5)	8	(3)	7	(2)	13	(5)
(193)	81 / 65	13	(5)	0	(1)	10	(4)	8	(3)
	88 / 83	13	(5)	5	(2)	3	(1)	23	(6)
T1'/T2'	14 / 12	15	(5.5)	0	(1)	25	(5.5)	5	(2.5)
(215)	66 / 16	10	(4)	15	(5.5)	5	(2.5)	5	(2.5)
	99 / 84	5	(2.5)	5	(2.5)	5	(2.5)	25	(5.5)
T1'/T2'	67 / 58	12	(3)	10	(2)	0	(1)	17	(5)
(232)	76 / 52	20	(5)	14	(4)	2	(2)	12	(4)
	96 / 61	30	(6)	0	(1)	20	(6)	10	(3)
T1"/T2"	78 / 68	6	(3)	0	(1)	1	(1)	5	(2)
(161)	/ 77			5	(2)			10	(3)
T1"/T2"	10 / 11	30	(6)	5	(3.5)	15	(5)	10	(2.5)
(215)	59 / 60	5	(3.5)	0	(1.5)	5	(1)	10	(2.5)
	70 / 95	10	(5)	0	(1.5)	15	(5)	15	(5)
T1"/T2"	71 / 62	49	(4)	0	(1)	30	(3)	8	(1)
(232)	87 / 73	67	(6)	17	(3)	102	(6)	52	(5)
	90 / 94	52	(5)	4	(2)	32	(4)	12	(2)
	mean *WTA*	19.4		4.6		17.0		14.5	
	median *WTA*	12.5		4.5		10.5		11.5	
	rank sum		(91.5)		(46.5)		(64.5)		(73.5)

Details of the Permutation Test

For the non-reference attribute, i.e. for columns 5 and 6 of Table 4.2, the test works as follows: *WTA* values are ranked for each *cell* of the table individually; ranks are assigned to observations from paired offer sequences (*T1'* compared to *T2'* or *T1"* compared to *T2"*) and from subjects that have the same monthly rent R_i as basis for the indifference equations. For WTA_{NRef} there are six such cells with six observations each and two cells with three observations each in Table 4.2. The permutation test creates all permutations of the 42 observations for WTA_{NRef} under the conditions that (1) each observation has to be assigned to its original cell and (2) the number of observations per treatment is not changed. This specific structure assures that, for example, data from *T1'* are not directly compared to *T2"* and that data from subjects with different final contracts are not directly compared. The purpose of this grouping is to reduce external variability. If one would, for example compare $WTA_{NRef}^{T1'}$ to $WTA_{NRef}^{T2"}$, one would compare the attribute balcony to the attribute elevator. A difference might arise from the external values of these issues rather than the offer sequences. Furthermore, comparing across different R_i could introduce wealth effects.

Under the side constraints for permutations, the observations are assumed to be independent and exchangeable, and hence, the test is unbiased, i.e. it is more likely to reject a false hypothesis than a true one (Good, 2000). The test's underlying idea is that the assignment of a subject to a cell is randomly determined by the experiment software. Within each cell the observed *WTA* values are compared to all possible other arrangements of these values to identify systematic differences between treatments. Overall, there are $N = \binom{6}{3}^6 \binom{3}{1}^2 = 576 * 10^6$ permutations. The rank sum of column 5 is the test statistic. For each permutation, the test statistic is computed and all possible values of the statistic taken together give the permutation distribution; it ranges from 40 to 94 with mean and median at 67.

Transferred to the test statistic, the null hypothesis of no treatment effect is equivalent to an equal rank sum for columns 5 and 6. Given the permutation distribution, the observed rank sum of 64.5 does not constitute an extremely low or high value (two-sided test, p-value $= \frac{395,858,896}{N} > 0.5$). Thus, the null hypothesis cannot be rejected.

Reference Attribute

For the reference attribute, the rational choice model predicts no treatment effect whereas the attachment effect model predicts greater values, and hence a greater rank sum, for *T1* than for *T2*. The null hypothesis tested in a one-sided rank-based exact permutation test (as described above) is that the rank sum in *T1* is less or equal to the rank sum in *T2*.

The rank sum of column 3 is the test statistic. The permutation distribution function for this test statistic ranges from 38.5 to 95.5 with mean and median again at 67. The slight difference of the distribution's range compared to the distribution function for the non-reference attribute comes from different ties in the data. Given this distribution, the observed rank sum of 91.5 can be identified as an extremely high value—a sum this high or higher occurs for only 820 permutations. Thus, the null hypothesis has to be rejected (one-sided test, p-value $= \frac{820}{N} < 0.001$). Hence, there is a significant difference between treatments as predicted by the behavioral model. The difference is statistically significant and, more importantly, it is of economic relevance: if the feature of an apartment—i.e. the elevator or the balcony—is included in every single of up to six offers by the agent, then on average the *WTA* value is more than four times as high as if the feature is not included in any single offer.

RESULT 4.2: *There is a substantial and statistically significant treatment effect for the attribute on which the agent's offer sequences are non-randomized (the reference attribute).*

Summary

Overall, the attachment effect model based on issue-wise reference points that change endogenously during a negotiation is in line with the data for the reference and the non-reference issue. On the other hand, the rational choice model of exogenously given and invariable preferences cannot account for the observed differences on the reference attribute.

RESULT 4.3: *The attachment effect model organizes data from the internet experiment on multi-issue negotiations better than the rational choice model does.*

4.3 Modifications of the Experiment Design

The treatment effect reported in the last section cannot solely be explained by random variation in the students' responses; instead, it suggests that the attachment effect model is significantly more accurate in predicting the subjects' behavior than the rational choice model is. Nevertheless, critiques might question the validity of the results on several grounds. As in any experiment, numerous design choices had to be made and each of them might be viewed skeptically. To lessen potential concerns—or put them aside completely, at best—, a second experiment was conducted.

Field experiments can help in designing better lab experiments. Thus, they have a methodological role quite apart from their complementarity at a substantive level (Harrison and List, 2004). In this context, lessons learned from the internet experiment are taken to design a follow-up lab experiment. The two designs are closely related— differences are pointed out in the following. Internal validity of the internet experiment can be challenged for the following aspects:

Control of environment: In an internet experiment, the experimenter has less control over the subjects' environment than in the lab. In the present experiment it was, e.g., not controlled whether a subject chatted with other persons during the negotiation, learned about others' experience with the experiment, or acquired information about the object of negotiation during the negotiation. Furthermore, it appears unlikely that subjects participated multiple times but this cannot be ruled out. These characteristics of the experiment can be seen as a strength concerning external validity or as a weakness with respect to internal validity (cf. Harrison and List, 2004).

To control the aforementioned factors, the second experiment is conducted in the lab.

Salient rewards: The internet experiment did not offer any salient reward, i.e. each participant had the same chance of winning the lottery independent of her actions during the experiment. It can be questioned whether subjects' were sufficiently motivated to sincerely consider the choices (although this raises the question what else could have caused the difference between treatments that was predicted by the behavioral model and observed in the data). In experimental psychology this question would hardly be a concern, in experimental economics it is highly debated (cf. Guala, 2005, Ch. 11).

To put concerns regarding salient rewards aside, the lab experiment uses salient rewards.

Sample size: The data set comprises 42 observations: 20 in the first and 22 in the second treatment. The test employed takes this sample size into account, and hence, the rejection of the null hypothesis for the reference attribute is already conditioned on this sample size. Nevertheless, a bigger sample might lead to more confidence in the results.

The lab experiment uses a sample about twice as big.

Besides the internet experiment's internal validity, its external validity might be questioned as well. Again, the second experiment alters several design features to reduce (or put aside completely) possible concerns.

Object of negotiation: In the internet experiment, subjects negotiated on the attributes of a (hypothetical) rent contract. *Multi-attribute* negotiations are one form of *multi-issue* negotiations, but not the only one. Other issues might, e.g., be the elements of a product bundle.

To test the different models in another environment than a multi-attribute negotiation, the lab experiment uses negotiations over product bundles.

Response modes: In Chapter 2, empirical evidence for preference reversals was presented—it was shown that revealed preferences oftentimes depend on the preference elicitation method. Thus, the question arises whether the preference elicitation used so far is the driving force behind the treatment effect.

The internet experiment elicited preferences via indifference equations, i.e. 'matching mode', where alternative-wise information processing can be expected (cf. Sec. 2.3.2). On the contrary, the lab experiment uses 'choice mode' for which issue-wise information processing appears more likely. Furthermore, several other measures (satisfaction rating, self reports on the complexity of choice, and response times) are used to assess subjects' preferences and the potential impact of reference point shifts.

Another rational for employing choice mode in the lab experiment is the prominence hypothesis: the more important issue has a higher impact in choice than in matching (cf. Sec. 2.1.2; Tversky, Sattath, and Slovic, 1988). If the negotiation process has an influence on which issue becomes the most important for a negotiator—and the internet experiment favors this assumption—then there might be a stronger treatment effect with choice mode rather than matching mode.

Sparse variation: The software agent in the internet experiment uses four different offer sequences. The randomization of the non-reference attribute and the monthly rent are thereby just made once. Hence, the observed effect might be an artifact of these very specific offer sequences.

To proof validity beyond these specific offer sequences, the lab experiment uses other sequences and more different sequences.

Nature of the counterparty: The fact that subjects negotiate with software agents rather than humans and the implications for generalizing to negotiations among two humans were discussed in Section 4.1.4. This feature is retained for the lab experiment, as the agents are a pivotal element of the experiment design to ensure internal validity.

Subject pool: All participants of the internet experiment were university students. Standard criticism to the external validity of student subject pools apply (cf. Friedman and Sunder, 1994, Ch. 4; Harrison and List, 2004). On the other hand several studies have presented evidence that students are not too unrepresentative for the overall population. The purpose of the experiment is not claiming that attachment effect applies in *every* negotiation for *all* negotiators but rather to say that it *might* occur for *some* negotiators. For this, a student sample is sufficient. Thus, it is retained for the second experiment. No subject participated in both experiments.[14]

Qualitative prediction: The hypotheses derived in Section 4.1.5 for the attachment effect model suggest the direction of a treatment effect. This qualitative prediction is supported by the data. The influence of single offers during the negotiation was, however, not quantified. To overcome this, the data from the second experiment will be used to estimate the parameters of a simplified version of the attachment effect model presented in Section 3.3. The larger sample size is important for the quality of this estimation.

To summarize, the lab experiment differs from the internet experiment in a variety of design decisions. It's not that one of the two experiments would be better than the other or an add-on to the other—both

---

[14] Furthermore, it might be expected that these specific subjects are less prone to preference changes as all of them (should have) attended lectures on microeconomics and should be aware of normative decision-making models. If so, finding a treatment effect for this subject pool might be an especially strong result. However, this reasoning is highly speculative.

are self-sufficient and valid on their own grounds. The combination of the two, however, transcends each single one in terms of internal and external validity as it demonstrates that the observed attachment effect is not closely linked to specific features of the design. The next chapter presents the lab experiment.

5

Laboratory Experiment

> *Finally, understanding decision requires knowledge beyond the traditional bounds of economics [...] The economic literature is not the best place to find new inspiration beyond these traditional technical methods of modeling.*
>
> (Smith, 2003, p. 510)

The internet experiment reported in the Chapter 4 supports the existence of an attachment effect in negotiation: Negotiators form issue-wise reference points during a negotiation and these reference points influence their preferences. To strengthen this result, a laboratory experiment is conducted. The major differences in the design of the two experiments are incentive compatibility, bundles of durable consumer goods instead of attributes of a hypothetical rent contract, a larger sample size, and several distinct measures for treatment effects (cf. Sec. 4.3).

The chapter is structured as follows: Firstly, Section 5.1 presents the experimental design and derives several measures and hypotheses to test for the attachment effect. Section 5.2 then introduces the statistics to be used in the *non-parametric analysis* of the data that is presented in Section 5.3. Subsequently, Section 5.4 estimates the parameters of the (linear) attachment effect model by means of the maximum likelihood method. This is the so called *parametric analysis* of the data. Finally, Section 5.5 summarizes the overall results that further support the existence of an attachment effect in negotiations.

5.1 Experimental Design

The experiment controls the course of an alternating offer multi-issue negotiation and uses a between-subject comparison of ex-post preferences. Control over the negotiation is achieved by determining the strategy of a software agent that takes the role of one negotiator and exchanges offers with a subject. This far, the experiment design is identical to the design of the internet experiment.[1] Ex-post measurement of preferences is based on a binary choice of each subject on which issue is most important for her.

Object of Negotiation

The negotiations are about bundles of three goods. Each subject can independently choose over which specific goods she wants to negotiate. Therefore she selects exactly one product from category 1 and exactly one product from category 2. Category 1 products are

- key cords with an imprinted university logo and
- espresso pots

and category 2 products are

- coffee mugs with an imprinted university logo and
- thermos flasks.

The third product in the bundle are recordable compact discs (CDs) with a capacity of 700 mega byte.

All products are readily available in shops close to the campus. Retail prices are about € 3.50 for a key cord, € 5.50 for an espresso pot, € 5.00 for a coffee mug, € 5.50 for a thermos flask, and € 0.40 for a recordable CD. The objective of this product selection is to increase the likelihood that the selected products have positive value for the subjects.

The agreement space is zero to two units of product 1, zero to two units of product 2, and zero to forty units of product 3. Each offer and each agreement has to be part of this agreement space. At the end of a session, these products are handed out to subjects. The number of units of a product a subject receives thereby depends on the contract she negotiated during the experiment. In addition, each subject receives a fixed amount of € 5. The usage of durable consumption goods as reward medium will be discussed in Section 5.1.4.

---

[1] Some aspects of the design overlap with the design of the internet experiment. Nevertheless, many of them are described in detail here in order to have a continuous description and to avoid various cross references.

Negotiation Protocol

The negotiation protocol studied are bilateral alternating offer multi-issue negotiations. Each subject negotiates with a software agent (termed *agent A*) independent of the other subjects. The agent starts with an initial offer and afterwards the negotiators take turns in making offers. An offer is a specification of the amount of different products in the bundle of three products, i.e. each offer consists of exactly three integers specifying a point in the agreement space. There is no possibility to exchange messages other than offers, like for example arguments, and no possibility to have a negotiation agenda—all issues are negotiated simultaneously.

The negotiation ends by either of two events: (1) one negotiator accepts an offer by the counterparty or (2) overall 12 offers are exchanged without any of them being accepted. In the former case, the accepted offer becomes the agreement between the subject and agent A. In the latter case, an arbitrator steps in and imposes a binding agreement. The arbitrator is implemented in the experimental software.

Instructions

Subjects are informed about the object of negotiation, the negotiation protocol, and the other rules of the experiment accurately and without deception. Most prominently, they are told that they negotiate with software agents. The instructions are handed out and read aloud at the beginning of the experiment.[2]

5.1.1 Procedure

Subject Pool

Subjects were students at the University of Karlsruhe. A random sample was selected from a database of students that voluntarily signed up to participate in economic experiments. Most subjects are undergraduate students of the School of Economics and Business Engineering. The participants' age, courses of studies, experience with experiments, etc. are detailed in Section 5.3.1.

Subjects were contacted via e-mail about a week prior to the session they were supposed to participate in. The e-mail offered several different dates and times for which the subject could sign up; it did not mention details on the experiment like, for example, that it would be

---

[2] The precise text is given in an appendix which is available upon request.

a negotiation experiment. Subjects were asked to indicate at which of the times offered they could participate. Sessions were then filled with the respondents and the dates were confirmed.

No subject participated more than once.

Experimental Technology

The experiment was conducted via computer terminals in a client-server architecture. The subjects' front end was a web browser (Microsoft Internet Explorer running on Microsoft Windows XP operating system) displaying HTML pages that include JavaScript to check subject entries before sending information to the server. As input device, subjects solely used a computer mouse.

On the server side, Java Servlets running on the Apache Tomcat Server handled the subjects' requests. All software agents were integrated in the servlets and their strategies were pre-specified by configuration files.

In addition, an administration screen allowed starting the experiment, to monitor progress, and to print out the final results. The software is an adaptation of the system programmed for the internet experiment (cf. Sec. 4.1).

Matching Procedures

Each subject participated independently of the other subjects, i.e. the decisions of one subject did not influence the information displayed to others, the offers presented to them, or the outcome for other subjects.

Payments

At the end of the experiment, subjects were paid € 5 cash and received the products that were specified in their final contract. Payments were made separately and in random order, so that no subject could infer the outcome of another subject.

Number of Subjects used in a Session

Due to no-shows, the number of subjects per session varies. Two sessions were conducted with 15 subjects each, another three sessions with 14 subjects, and the final session with 12 subjects. Overall, this amounts to 84 subjects. However, data from 2 participants in the last session is dismissed as the data analysis bases on paired observations. The decision to dismiss data from exactly these subjects was made automatically by the software. The decision was independent of the subjects' behavior in the negotiation and subsequent choices; the details of this selection process are explained in Section 5.3.1, the entire data is given in an appendix which is available upon request.

Timing

The experiment was conducted in December 2005 (sessions 1 to 3) and January 2006 (sessions 4 to 6) in the laboratory of the Institute of Information Systems and Management, University of Karlsruhe. A typical session lasted about 45 minutes. Thereof, 15 minutes were instructional.

5.1.2 Course of a Session

When showing up for the experiment, subjects were asked to wait in the hallway in front of the lab. At the time the experiment was scheduled they entered the lab, randomly drew a sheet of paper that assigned them to a computer terminal, and took a seat in front of the respective terminal. All terminals were visually separated from one another. At her place, each subject had a computer monitor and mouse, a printout of the instructions, a blank sheet of paper, a voluntary feedback form asking for comments on the experiment and the individual strategy, a receipt for the payment, a pencil, and a ruler.[3]

Instructions

The four pages of instructions were read aloud by a hypothesis-blind research assistant. The main topics are (1) a general introduction, (2) an explanation of the abstract task of negotiating a product bundle, (3) a description of the specific products, (4) the procedure of negotiating with software agents (see below), (5) payment conditions, and (6) a generic ending that instructs subjects how to proceed in case they have a question. At the point that describes the products, reading the instructions was interrupted for a while and two research assistants presented samples of the products to subjects.

Questionnaire

Subsequent to the instructions, participants were required to answer 14 multiple-choice questions on the rules of the experiment to assure comprehension of the rules. The system responded to incorrect answers by asking to look up the correct answer in the instructions. The same

---

[3] Copies of the documents are provided in an appendix which is available upon request.

question was asked over and over again until it was answered correctly. The single questions and possible answers are provided in an appendix which is available upon request.

Product selection

The first choice a subject has to make in the experiment is over which specific products she wants to negotiate. For product 1, she can choose between key cords and espresso pots, for product 2 between coffee mugs and thermos flasks, and product 3 are recordable CDs for all participants.

The selection of a product 1 and a product 2 is entered on the same screen. Photos of all products are shown to increase the subjects' attachment to these products. On the following screens, just the photos of the two selected products and the CDs are shown. The subjects' screen is shown in Figure 5.1.[4]

Agent A

The protocol of the negotiation with agent A is the alternating offer exchange described at the beginning of Section 5.1. Each time the agent makes an offer, the subject can either accept this offer by clicking on one button, or propose a counteroffer by first specifying it via selection from three drop-down boxes and then clicking a 'send counteroffer' button. Additionally, the history of all previous offers is provided at the lower end of the screen; see Figure 5.2 for a screen shot.

After the negotiation terminated—either because an offer was accepted, or because the maximum of 12 offers was reached—subjects are informed about the outcome, i.e. their agreement with A, and are asked to specify their satisfaction with each single attribute as well as with the entire agreement. Each of the four satisfaction ratings is entered on a separate five-step Likert scale ranging from 'very dissatisfied' to 'very satisfied'. A screen shot of the satisfaction rating is given in Figure 5.3.

Agent B

Agent B makes one take-it-or-leave-it offer. The agreement with agent A and the offer by agent B are displayed side by side and the subject can choose either of the two. If she chooses the agreement with agent A, this remains her agreement. If she chooses the offer by B, the agreement with A is resolved and the subject has a new agreement with agent B. The subjects' screen for this task is given in Figure 5.4

---

[4] This and all following screen shots are translated to English; original screen shots are provided in an appendix which is available upon request.

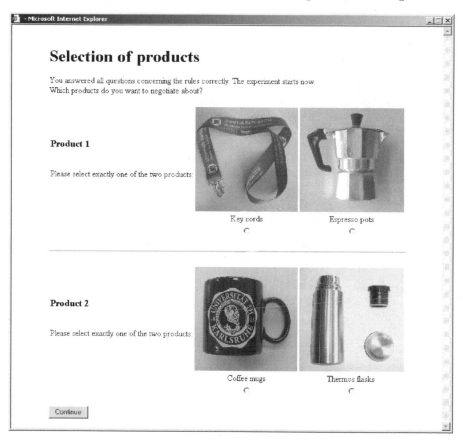

Fig. 5.1. Product selection (screen shot)

In addition to choosing either the agreement with A or the offer by B, each participant is asked to rate the complexity of this choice on a five-step Likert scale with labels 'very easy' on the left and 'very difficult' on the right side.

By design, the offer of agent B dominates the agreement with agent A (this relation will become clear in the following Section 5.1.3 that outlines the agents' strategies). Thus, it is expected that all subjects accept B's offer. Furthermore, the offer by B is the same for all subjects in the experiment—the purpose is to level all subjects on the same agreement irrespective of their performance in the negotiation with agent A to rule out wealth effects.

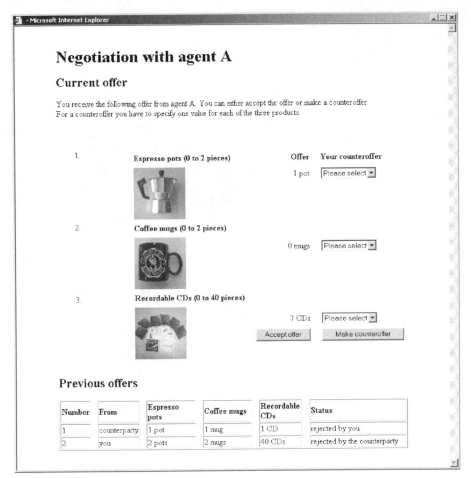

Fig. 5.2. Alternating offer negotiation (screen shot)

Agent C

Subsequently, agent C makes another two offers. Again, these are the same for all subjects: The first offer gives one unit of the first issue and zero on the second and third issue. The second offer gives one unit of the second issue and zero units of the first and third. The subject has to choose either of the two offers—the corresponding screen shot is given in Figure 5.5.

As both offers grant one unit of one issue, choosing an offer is essentially the same as choosing the first or the second product, i.e. it is a revelation of preferences for these two products. This choice is the most

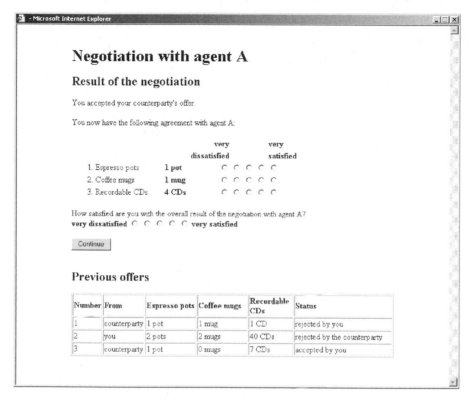

Fig. 5.3. Satisfaction rating (screen shot)

central measure on whether agent A had an influence on preferences or not.

In addition to choosing either offer, the subject is asked to rate the complexity of this choice, again on a five-step Likert scale.

Overall Result

A participant's payment is determined by the toss of a virtual coin: with a 50% chance the subject's payment is the agreement with either agent A or B (whichever the subject chose) and with a 50% chance it is the agreement with agent C. The result is termed the subject's contract and displayed on the screen. This random draw creates two essential properties of the overall procedure:

1. The agreements with agents A, B (if the offer is accepted), and C can all potentially determine a subject's payoff. Thus, the subject has an incentive to do 'as good as possible' in the interaction with

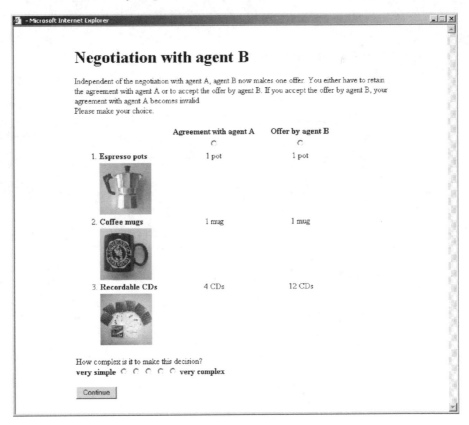

Fig. 5.4. Offer by agent B (screen shot)

every single agent. This especially means negotiating eagerly with agent A and choosing the preferred product from agent C.

2. In case the choice between offers by agent C has an impact on a subject's payment, the agreement with agent A or B has not. Thus, preferences over the products offered by C should be independent of the agreement with A or B. Reasoning on substitutes, complements, or (diminishing) marginal utility like 'I got an espresso pot from agent A, so now I choose the thermos flask from agent C as I do not need a second espresso pot' is irrelevant.

Exit Questions

The final screen allowing subject input asks for the course of studies, age, gender, number of laboratory experiments the subject had already participated in, and the number of negotiation experiments thereof. Answering the questions was voluntary.

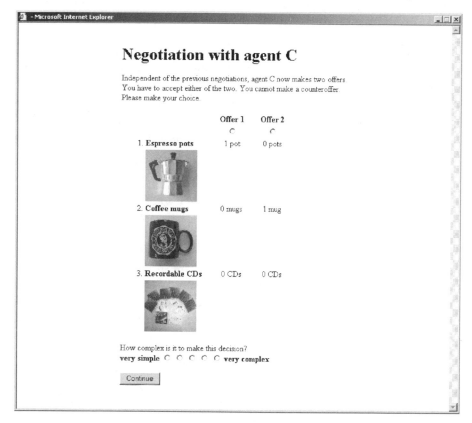

Fig. 5.5. Offers by agent C (screen shot)

Payment

Subjects were paid in random order and separate from one another after all subjects were finished. They had to fill out a receipt for the payment and had the possibility to return the feedback form if they liked.

The entire procedure (except the specific offers by the agents) was explained to subjects in the instructions. The instructions and questions on their understanding especially pronounced that (1) all subjects participated independently of each other, (2) subjects negotiated with software agents, (3) the three software agents and the arbitrator acted independently of each other without knowledge of the subject's previous actions, and (4) the random draw at the end.

5.1.3 Treatments and Agent Strategies

The experimental procedure involves three different software agents interacting with the subjects (plus eventually the arbitrator). On a conceptual level, their primary purpose can be classified as follows: Agent A manipulates the subjects' reference points, agent B levels all subjects in terms of expected payment to rule out wealth effects, and agent C measures ex-post preferences. The offers and strategies employed to do so are outlined in the following.

Treatments

The experiment involves two treatments: treatment 1 aims at creating a high reference point on issue 1 and treatment 2 on issue 2. Consequently, issue 1 is termed *reference issue* for treatment 1 and *non-reference issue* for treatment 2. Analogously, issue 2 is the non-reference issue in treatment 1 and the reference issue in treatment 2. The only treatment variable in this experiment is the usage of either issue 1 or issue 2 as reference issue. The terms *reference product* and *non-reference product* are employed as synonyms for the respective issues, as the issues under negotiation are products in this experiment.

To achieve a reference point, all offers of agent A grant a constant value of one unit on the respective reference issue. For the non-reference issue and issue 3 (the CDs), the values are randomized to create a sequence of offers the agent proposes to its counterparty.

One could think about a third treatment manipulating the reference point on issue 3. However, this would require increasing the sample size without allowing a substantially different analysis of the hypotheses. The reason to include issue 3 is rather that the relatively low value of a single recordable CD (retail price € 0.40) allowed to increase the agreement space to overall $3 \times 3 \times 41 = 369$ different possible offers and agreements. This is important to allow for meaningful differentiation of offers in the negotiation with agent A. Basically, the CDs serve as numeraire where money would be to prominent an issue.

Offer Sequences

For each subject, agent A has a predefined sequence of six offers it will propose in case the negotiation does not end earlier. Three examples of such sequences are displayed in Figure 5.6. The horizontal axis gives the number of units of product 1, the vertical axis the units of product 2. Single offers by agent A are given in parenthesis with firstly the time and secondly the number of units of product 3. Thus, in Figure 5.6a, for example, agent A starts the negotiation with an initial offer of one unit of each product (denoted as $\langle 1, 1, 1 \rangle$ in the text) at time $t = 1$. At time $t = 2$, the subject might counter with another offer and if the agent should make a new offer at time $t = 3$, this will be $\langle 1, 0, 7 \rangle$, etc. The dimensions are discrete, i.e. the specific location within a 'cell' in Figure 5.6a is just for layout reasons and has no further meaning. Thus, the offers at time $t = 3$ and time $t = 5$, for example, are equal with respect to product 2.

Figure 5.6a shows the sequence of offers that was presented to subject 1 and Figure 5.6b the sequence for subject 2. Several such sequences of offers were determined randomly for treatment 1 prior to the experiment with the following side conditions: Firstly, they never offer more than one unit of either product 1 or product 2, and never more than twelve units of product 3. Thus, agent A only uses a subspace of the possible agreement space. Secondly, each sequence starts with an offer that is expected to be quite undesirable for the subject and subsequently makes small concessions to the subject or trade-offs most of the times. Thirdly, and most importantly, every single offer includes a fix amount of one unit of product 1, i.e. the reference product in treatment 1.

For agent A in treatment 2, the strategies are derived by taking the strategies from treatment 1 and exchanging the labels of issues 1 and 2. Figure 5.6c shows, for example, a sequence of offers in treatment 2. This sequence corresponds to the offer sequence from treatment 1 that is displayed in Figure 5.6b. The only difference is that offers from treatment 1 with one unit of issue 1 and zero units of issue 2 become offers in treatment 2 with zero units of issue 1 and one unit of issue 2. The agent's last offer ($t = 11$), for example, is $\langle 1, 0, 7 \rangle$ in treatment 1 and $\langle 0, 1, 7 \rangle$ in treatment 2. Offers in the two treatments are paired to increase comparability between treatments.

All offer sequences used can be found in an appendix which is available upon request.

Assignment of Subjects

Subjects are randomly assigned to treatments by the experiment software. Pairs of subjects in different treatments that selected the same combination of products (e.g. espresso pot / coffee mug) are assigned corresponding strategies by agent A, i.e. strategies that just differ by exchanging the labels of issues 1 and 2. Within a single treatment, each offer sequence is used just once. Thus, data from subjects is paired across treatments but nevertheless observations are independent of one

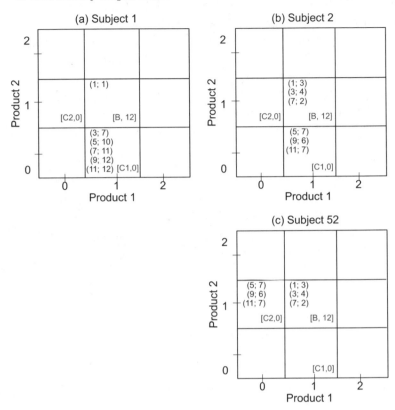

Fig. 5.6. Examples for offers by the software agents

another, as the actions of one subject do not influence the other participants.

Strategy of Agent A

The strategy of agent A is straightforward: it is endowed with a sequence of six offers and sends one after another to the subject. This process ends with either of three events: (1) the subject accepts one of the agent's offers, (2) the agent accepts the subject's counteroffer, or (3) 12 offers are exchanged and the arbitrator imposes an agreement.

The agent's acceptance policy is as follows: reject any offer by the subject except it weakly dominates either a previous offer by the agent or the next offer to be made. Dominance is seen from the agent's perspective here, i.e. for each issue individually the subject's offer asks for at most as many units as the agent's offer grants. Formalized, the policy reads as follows: Given the subject's offer $x^t \in X$ and the sequence of agent offers $x^1, x^3, \cdots, x^{t+1} \in X$, the agent accepts if and only if $\exists\, t' \in \{1, 3, \cdots, t+1\} : x_k^t \leq x_k^{t'} \forall\, k \in \{1, 2, 3\}$.[5]

The rational for this policy is that a subject might be irritated if, for example, the agent rejects her offer and counters with exactly the same offer.

Arbitrator

The agreement the arbitrator imposes in case the negotiators do not reach an agreement is randomized prior to the experiment for each pair of subjects. As a side condition, it is not better for the subject than the offer by agent B.

Strategy of Agent B

Agent B just makes a single offer and this is the same for all subjects. Agent B offers $\langle 1, 1, 12 \rangle$; the offer is displayed as '[B,12]' in Figure 5.6. Given the side conditions for generating offer sequences as outlined above and agent A's acceptance policy, agent B's offer by design (weakly) dominates any agreement with agent A the subject might have. (Note that weak dominance is seen from the subject's perspective here, i.e. the offer by B is at least as good on every issue as the agreement with A.)

The purpose of introducing agent B is that all participants accept this offer and have the same expected outcome so far when coming to the choice among offers by agent C. Otherwise, differences in the choice of one of C's offers might be due to wealth effects.

Strategy of Agent C

Agent C simply offers the same two product bundles to any subject. The first bundle contains just one unit of issue 1 ($\langle 1, 0, 0 \rangle$) and the second bundle offered includes solely one unit of issue 2 ($\langle 0, 1, 0 \rangle$). The two offers are displayed as '[C1,12]' and '[C2,12]' in Figure 5.6. These two offers are a simple way to elicit a subject's preferences with respect to whether she prefers product 1 over product 2 or the other way round. In case the subject would be indifferent, she has to resolve the conflict and choose either offer.

---

[5] An additional offer $x^{13} = \langle 0, 0, 0 \rangle$ is added to the agent's offer sequence to allow the computation of whether to accept the final offer in the negotiation. This last (technical) offer is never send to a subject.

5.1.4 Discussion of Design Features

The experimental design outlined so far has two features that might appear unconventional: The first feature are the durable consumer goods as reward medium instead of money and the second feature is the play against software agents instead of human opponents. In the following, the reward medium is reviewed critically; see Section 4.1.4 on the internet experiment for a discussion of the play against software agents. There it was concluded that the usage of software agents on the one hand increases the experimenter's control over the negotiations but that, on the other hand, it limits external validity.

Induced Values

Many economic experiments base on induced-values: The key idea is, as described by Friedman and Sunder (1994, Ch. 2.3), to induce specific characteristics to subjects so that their innate characteristics become largely irrelevant. This allows the researcher to gain control over, e.g., the subjects' risk attitude, time discounting, or the relative weighting of attributes in a multi-attribute decision-making experiment.

Smith (1976b), Wilde (1980), and Smith (1982a) propose a set of sufficient conditions—so called precepts—to gain control over preferences in a laboratory experiment: *non-satiation, salience, dominance, privacy,* and *parallelism.* Note however that these are sufficient conditions, not only necessary conditions. Non-satiation of the reward medium requires monotone increasing utility over the amount of the medium, i.e. everything else being equal the subject prefers more of the reward medium to less of it. Salience means that the reward a subject receives depends on her actions—a fixed show up fee is non-salient, rewards linked to behavior are. Dominance requires that changes in a subject's utility from the experiment should predominantly come from the reward medium and not from other influences. Privacy requires that each subject in the experiment only obtains information on her own payoffs to rule out interpersonal utility considerations.[6] Parallelism, finally, means that propositions derived from the lab apply also to non-laboratory settings where similar ceteris paribus conditions hold. What precisely 'similar ceteris paribus conditions' are cannot be defined on a general basis. Parallelism is oftentimes referred to as *external validity.* While the first

---

[6] Note that in recent years several experiments have approached topics like fairness, positional goods, and other aspects of other-regarding preferences. Here the privacy of payoffs (though not the anonymity) is purposefully relaxed to explicitly allow interpersonal utility considerations.

four precepts directly refer to the subjects' incentives, external validity does not.

Local currency is oftentimes chosen as reward medium to induce preferences over abstract outcomes that closely resemble the abstract and mathematical nature of game theoretic equilibrium models. Monetary incentives can be linked to numerous abstract settings and they are convenient to implement. However, money is just one possible reward medium: Smith (1982a, p. 931) points out that '[...] in the laboratory we also have to induce value on outcomes with a monetary (or other) reward function.'[7] The consumer goods used in the present experiment fall in the category of Smith's 'other reward function'.

Monetary incentives and consumer goods both satisfy the first three requirements outlined above: non-satiation,[8] salience, and dominance. Privacy depends on the information structure during and after the experiment. As subjects are unaware of their fellow subjects' performance and payments are made separately, privacy is achieved as well. Parallelism—the last condition set out by Smith (1982a)—is the factor in which money and consumer goods as two possible reward media differ. Results derived with one or the other (might) belong to different ceteris paribus conditions and (might) differ with respect to their external validity.

Abstraction

The abstraction used in 'context-free' game theoretic laboratory experiments is oftentimes seen as a virtue of this methodology as confounding factors are ruled out to a wide degree. Eckel and Grossman (1996, p. 188), for example, note that it 'is received wisdom in experimental economics that abstraction is important [and] experimental procedures should be as context-free as possible [...]'. In their next paragraph, the same authors highlight that 'Economists are becoming increasingly

---

[7] In fact, Smith (1982a, p. 931) devotes an almost half-page long footnote to arguing that monetary rewards are almost the same thing as having 'real preferences' which is seen as preferable by some critiques.

[8] It might be questioned whether subjects have positive marginal utility for a second key cord, espresso pot, coffee mug, or thermos flask. (Negative marginal utility would be implausible as subjects could refuse to receive the second unit of a good.) However, firstly by design no subject obtains two units of either of these goods but at most one, secondly the product selection at the beginning serves to increase the likelihood of positive utility, and thirdly the observed negotiation behavior with numerous subject offers demanding two units of the goods, ex-post allows to conclude that at least most subjects had positive marginal utility for the second unit.

aware that social and psychological factors affect economic decision-making, and the importance of social factors can only be introduced by abandoning, at least to some extent, abstraction' (p. 189). This argumentation is in line with the methodological discussion of field experiments recently presented by Harrison and List (2004): the abstraction used in many laboratory experiments might be interpreted as control, on the other hand it might as well sacrifice control as the researcher can no longer influence which field context the subjects' bring to the abstract laboratory game. Thus, these authors argue that abstract settings with monetary rewards might not resemble 'similar ceteris paribus conditions' of any meaningful naturally occurring environment to which results might be generalized.

Partially abandoning abstraction and using field goods with the subjects' 'real preferences' (in the terminology of Smith, 1982a) is a pivotal element of the experiment design. Kahneman, Knetsch, and Thaler (1990) were among the first to systematically test the endowment effect on a large scale basis. They found substantial evidence for the endowment effect in experiments with durable consumer goods (e.g., coffee mugs) but no effect for abstract goods with induced monetary values. This corroborates the supposition of Heiner (1985) and Harrison and List (2004) that results from preferences induced via a monetary reward function might not generalize to other settings, e.g. because the abstract framing rules out psychological effects that occur with field goods. Another interpretation of the results by Kahneman et al. is based on expectations (Köszegi and Rabin, 2006): subjects do not expect to keep the money they are earning in an experiment for an extended period of time. Money is used as exchange medium on a regular basis and, thus, the subjects do not become attached to money. On the other hand, subjects might expect to retain the durable consumer goods they earn in an experiment for a longer time. Thus, they become attached to them and their preferences change via a shift of the reference point. Hence, using consumption goods as reward in the present experiment might be a necessary condition for a treatment effect to occur.

Types of Experiments

In a recent book on the methodology and philosophy of experimental economics, Guala (2005, Ch. 11) points out in line with Cubitt, Starmer, and Sudgen (2001) that economic experiments can mainly be classified in two distinct categories:

1. experiments aimed at testing the effect of individual preferences, beliefs, endowments, institutions, etc. on market outcomes
2. experiments aimed at testing the standard assumptions imposed on individual preferences, beliefs, etc.

For experiments of the first type, it is inevitable to try to implement the standard assumptions by inducing preferences that are consistent with the assumptions of rational choice theory. In experiment of the second type, on the contrary, the assumptions of decision theory are tested themselves—'the aim is to figure out whether individual preferences [...] have the structure postulated by the standard models.' (Guala, 2005, p. 236)

Smith's early experiments on price theory and market institutions certainly fall in the first category of experiments (Smith, 1976a, 1982b). These are the kind of experiments that the precepts outlined above were proposed for (Smith, 1976b, 1982a) and in these kind of experiments money is commonly used as reward medium—although not exclusively. Smith (1976b, p. 275) points out that his postulations for inducing values apply 'to experiments designed to test price theory propositions conditional upon known valuations. Separate experiments can be designed to test propositions in preference theory.'

The second category comprises experiments designed to test propositions in standard economic models. Here, other reward media are employed on a regular basis. The present experiment belongs to this second category.

Related Work

Field goods rather than abstract goods linked to monetary incentives have been employed in numerous experiments belonging to either of the above types of economic experiments. In the terminology proposed by Harrison and List (2004), such experiments are *framed experiments*. Among the experiments using field goods to test the implications of game theoretic models is the one by Bohm (1972) who used a closed-circuit broadcast of a new TV program (that was really produced and delivered to the subjects) to study elicitation of valuations for public goods; Brookshire and Coursey (1987) used an increase in density of park trees as experimental public good, and Cummings, Harrison, and Rutström (1995) as well as Rutström (1998) sold gourmet chocolate truffles to their subjects in a laboratory experiment on incentive compatible auctions. Bateman et al. (1997) used pizza and dessert vouchers as goods and Lucking-Reiley (1999) conducted experiments on revenue equivalence and reserve prices in internet auctions using collectible trading cards (see also Reiley, 2006). Finally, Hossain and Morgan (2006) sold CDs and Xbox games to test revenue equivalence.

In experiments challenging the assumptions of rational choice theory, the use of field rather than abstract goods is common. In their seminal article on prospect theory Kahneman and Tversky (1979), for example, used lotteries over monetary outcomes as well as over vouchers for holiday trips to demonstrate systematic deviations from the assumptions of expected utility theory. To test for the endowment effect, Knetsch (1989) offered his subjects the choice between coffee mugs and chocolate bars and Kahneman, Knetsch, and Thaler (1990) used coffee mugs, pens, binoculars, and chocolate bars. These goods are probably the most commonly used goods on the stability of preferences: coffee mugs were as well used by, for example, Loewenstein and Issacharoff (1994), Shogren, Shin, Hayes, and Kliebenstein (1994), Franciosi, Kujal, Michelitsch, Smith, and Deng (1996), Morrison (1997), and Arlen, Spitzer, and Talley (2002). Recently, sports cards have been used as commodity in several experiments on auctions or on the endowment effect (List and Lucking-Reiley, 2000; List, 2001, 2003, 2004; Harrison, List, and Towe, forthcoming). Some further experiments using field goods were cited in Chapter 2.

Summary

The present experiment intends to test an assumption that is inherent in rational choice models: preferences are invariable during a negotiation. In the design outlined above, the use of consumer goods plays a pivotal role. Favoring them over monetary incentives is guided by the following arguments: (1) they satisfy the requirements for a reward medium in experimental economics (Smith, 1982a; Friedman and Sunder, 1994), (2) they allow for subjects' uncertainty over their precise valuation for the rewards in a natural way, (3) they are not uncommon, and (4) they might be a necessary condition for a treatment effect to occur.

5.1.5 Hypotheses on Treatment Effects

The previous sections introduced and discussed the experiment design. The following now details the differences the treatments might create with respect to the subjects' preferences and derives testable hypotheses.

Models on a Negotiator's Preferences

The attachment effect model on the emergence of reference points in negotiations was introduced in Section 3.3. For the non-parametric analysis of the experiment, i.e. for Section 5.3, two different versions are used. The first version is termed *rational choice model*, the second *attachment effect model*. Strictly speaking, the rational choice model is a special case of the attachment effect model introduced in Section 3.3. However, here the term is loosely used for instances of the attachment effect model except the rational choice case.

The microeconomic rational choice model is a simplification of the attachment effect model where the initial reference point is any arbitrary but fixed point in the agreement space (if it exists at all) and all update functions are equivalent to zero ($f_k^+(\cdot) = 0$, $f_k^-(\cdot) = 0$, $g_k^+(\cdot) = 0$, $g_k^-(\cdot) = 0 \ \forall \ k \in \{1,2,3\}$). Thus, there is no change of a reference point during the negotiation. This setup is common (and implicit) in virtually all microeconomic models on rational choice.

For the attachment effect model on the other hand, it is assumed that a subject's reference point gradually moves towards the value offered by the agent ($\exists \, a > 0 : f_k^+(a) > 0$ and $\exists \, a > 0 : f_k^-(a) > 0 \ \forall \ k \in \{1,2,3\}$). No special assumption on the initial reference point or the update concerning the subject's own offers are made.[9] For the parametric analysis of the data as presented in Section 5.4, the full model (with linear update functions) is taken and the parameters are estimated.

By design of the offer sequences, one unit of the reference issue is included in every single offer whereas zero or one units of the non-reference issue are included. Thus, the reference issue is offered as least as often as the non-reference issue by agent A. If a subjects' reference point gradually moves towards the agent's offers—as it is assumed in the attachment effect model—then subjects tend to have a higher reference point on the reference issue than the non-reference issue. Subsequent to the negotiation with agent A, agent B's offer likely influences a subject's reference point as well. As this does not differ between either treatments or issues 1 and, this potential influence of agent B is neglected.

---

[9] If update functions and initial reference points are subject- and issue-specific, the very unlikely case could occur that for each single subject the sequence of offers exchanged with agent A incidently results in an ex-post reference point that equals the initial one. However, this extreme and unlikely case is neglected here. If it would occur, it would be indistinguishable from the rational choice model.

Binary Choice

The choice of either offer by agent C is the most straightforward revelation of a subject's preferences in the experiment.

It is assumed that prior to the experiment, each subject has *a-priori preferences* over the different goods. Let $\pi^0_\succ$ denote the probability that a single participant that is drawn from the subject pool a-priori prefers product 1 over product 2 ($P_1 \succ P_2$), $\pi^0_\sim$ the probability of $P_1 \sim P_2$, and $\pi^0_\prec$ the probability of $P_1 \prec P_2$. With complete preferences for any subject, $\pi^0_\succ + \pi^0_\sim + \pi^0_\prec = 1$ holds. Subjects are assigned randomly to treatments. Thus, the probabilities are a-priori the same for subjects in both treatments.

In the rational choice model, preferences are fix—when subjects choose among offers by agent C, preferences are the same as they were at the beginning of the experiment. Thus, when a single subject is randomly drawn, the probabilities for her different possible preferences are the a-priori probabilities independent of the treatment: $\pi^{T1}_\succ = \pi^{T2}_\succ = \pi^0_\succ$ where the superscript 0 stands for the a-priori probabilities and $T1$ and $T2$ for the ones when choosing among offers by C in treatment 1 and treatment 2, respectively.

In the attachment effect model, it is assumed that subjects tend to have a higher reference point on the reference issue than the non-reference issue. This corresponds to a shift of the reference point as it was exemplified in Figure 2.7 in analogy to Tversky and Kahneman (1991). A loss on the reference issue thus tends to loom larger than a loss on the non-reference issue. Hence, subjects who a-priori preferred the reference issue anyways will continue to do so, subjects who were indifferent prefer the reference issue after having negotiated with agent A, and potentially some of the subjects that a-priori preferred the non-reference issue changed their mind via repeated offerings by agent A and ex-post prefer the reference issue. As the reference issue differs between treatments, this is reflected in the likelihood of randomly drawing a single subject who prefers product 1 over product 2: $\pi^{T1}_\succ > \pi^0_\succ > \pi^{T2}_\succ$.

Under the assumption that the subject accepts the offer by agent B, the choice among offers by C is the choice between two lotteries concerning the subject's payment: The first lottery yields either outcome $\langle 1, 1, 12 \rangle$ or $\langle 1, 0, 0 \rangle$ with equal probability. The second lottery gives a 50% chance to each of the two outcomes $\langle 1, 1, 12 \rangle$ or $\langle 0, 1, 0 \rangle$. After eliminating the common consequences, i.e. the agreement with agent B, the subject's choice reduces to selecting either $\langle 1, 0, 0 \rangle$ or $\langle 0, 1, 0 \rangle$. Which of the two a subject selects, depends on her ex-post preferences. Thus, the frequency $p^{T1}_\succ$ of subjects choosing product 1 in treatment 1 compared to the frequency $p_{\succ}^{T2}$ in treatment 2 allows to draw conclusions on the relationship of $\pi_{\succ}^{T1}$ and $\pi_{\succ}^{T2}$—rational choice predicts equality and the attachment effect model a larger value for $\pi_{\succ}^{T1}$ than for $\pi_{\succ}^{T2}$.

Sources of Utility

The two sources of utility considered are *consumption utility* and *gain/loss utility*—overall utility is the sum of both utility sources (Köszegi and Rabin, 2006). At this point, utility is assumed to be a cardinal multi-issue measure (cf. Sec. 2.1.1). A subject's self rating of her satisfaction with each single issue of the overall agreement she negotiated with agent A serves as proxy for utility. This discrete measure on a five-step scale is a rather crude approximation of utility. However, it is seen as the least invasive of several possible utility elicitation techniques: leading the subject trough a complex procedure like SMART (Edwards, 1977), SMARTER (Edwards and Barron, 1994), AHP (Saaty, 1980), or the like might on the one hand distract the subject and on the other hand influence her preferences.

Consumption Utility

Consumption utility is the standard source of utility in microeconomic models: a subject is expected to have higher utility from a higher outcome. The rational choice model and the attachment effect model coincide in this. Monotone preferences on each issue individually and, thus, non-satiation of the reward medium are assumed. Furthermore, differentiability of the utility function is assumed. Then formally, the inequality $m_k(0) < m_k(1)$ is expected to hold for $k \in \{1, 2\}$ where m_k is the partial derivative of a subject's consumption utility function with respect to issue k.

Two aspects of this formalization are noteworthy: Firstly, it is assumed that preferences are monotone on each issue individually, i.e., for example, for any fixed number of units of products 2 and 3, a subject prefers one unit of product 1 to zero units of product 1. Additive utility functions are the most widely used heuristic for trading off multiple issues. The property of issue-wise monotone preferences holds for any such additive multi-issue utility function. Thus, additivity is a sufficient condition for the above inequality; it is, however, not a necessary condition here. (See Keeney and Raiffa (1993, Ch. 3, 5 & 6) for an extensive discussion of independence of issues and different forms of multi-issue utility functions.)

Secondly, and more generally, the inequality should hold for all sets of two outcomes where the first is greater than the second and for all three issues. However, for the sake of simplicity and to limit the scope of the assumptions this is omitted here, as just outcomes of zero or one unit on either issue 1 or issue 2 are analyzed.

Gain/Loss Utility

In the attachment effect model, gain/loss utility is the utility derived from an outcome given a reference point. The partial derivative with respect to issue k is denoted as $n_k(x_k \mid r_k)$ where x_k is the outcome and r_k the reference point on issue k. Gain/loss utility is assumed to be a function of the difference of outcome and reference point, to be strictly monotonically increasing in this difference, and to be zero for equality of outcome and reference point. These assumptions are common for prospect theoretic value functions like, e.g., the two-part power function proposed by Tversky and Kahneman (1992, cf. Sec. 2.2.1). Furthermore, loss aversion implies that losses loom larger than gains; this additional assumption is, however, not necessary at this point.

The rational choice model does not include the notion of gain/loss utility. Even with a fixed reference point in the model, the distinction of different sources of utility is irrelevant, as the consumption utility function can capture a subject's preferences with respect to such a fixed reference point.

Overall Utility

Overall utility is the combination of consumption utility and gain/loss utility (Köszegi and Rabin, 2006). Let $u_k(x_k \mid r_k) = m_k(x_k) + n_k(x_k \mid r_k)$ be the partial derivative of the overall utility function with respect to issue k. Given two different reference points r_k^R and r_k^N with $r_k^R > r_k^N$, the same outcome gives the same consumption utility but different gain/loss utility ($n_k(x_k \mid r_k^R) < n_k(x_k \mid r_k^N)$), and hence, different overall utility: $u_k(x_k \mid r_k^R) < u_k(x_k \mid r_k^N)$. Furthermore, given the two different outcomes of zero or one unit on issue $k \in \{1, 2, 3\}$, the same reference point implies different overall utility: $u_k(0 \mid r_k) < u_k(1 \mid r_k)$. Thus, given a single reference point, the subject get's higher utility from obtaining more. This standard assumption of consumption utility remains valid even if gain/loss utility is added as second source of utility.

Satisfaction

Utility is not observed directly but via participants' reported subjective satisfaction rating. Let $s_k(a) \in \{1, 2, 3, 4, 5\}$ be a subject's reported satisfaction with an outcome of $a \in \{0, 1\}$ units for issue $k \in \{1, 2\}$. Then the above inequalities imply $s_1(0) < s_1(1)$ and $s_2(0) < s_2(1)$. The rational choice model and the attachment effect model coincide in this supposition on consumption utility.

In the attachment effect model, it is assumed that subjects tend to have a higher reference point on their reference issue than on their non-reference issue. Thus, the above inequalities derived from the attachment effect model imply $s_1^R(a_1) < s_1^N(a_1)$ and $s_2^R(a_2) < s_2^N(a_2)$ where the subscript stands for issue 1 or issue 2, the superscript indicates whether this issue is the respective subject's reference issue (R) or not (N), and $a_1, a_2 \in \{0, 1\}$ is the subject's agreement with agent A on that issue. The intuition is that a subject gains higher utility from an unexpectedly good outcome than from an expectedly good outcome. The rational choice model, on the other hand, predicts no influence of the reference issue, i.e. $s_1^R(a_1) = s_1^N(a_1)$ and $s_2^R(a_2) = s_2^N(a_2)$.

Preference Uncertainty

Preference uncertainty means that one is unsure about one's preferences: either about the question which of several options one prefers, or how strongly one prefers one option over another (cf. Sec. 2.3; Fischer, Luce, and Jia, 2000; Fischer, Jia, and Luce, 2000). The former kind of preference uncertainty is analyzed in the choice among offers by agent C.

A-Priori Utility Differences

Under the assumption of cardinal multi-issue utility, preference uncertainty is oftentimes modelled by a random additive error term. At the beginning of the experiment, a subject would derive utility from obtaining a product bundle $x \in X : u(x \mid r^0) = m(x) + n(x \mid r^0) + \epsilon$ where ϵ is an additive random error term sampled from some distribution that does not depend on either the outcome or the reference point and has mean zero. Note that this random error term would not change the derivation of hypothesis with respect to consumption utility and gain/loss utility—above it was omitted for simplicity. See Fischer, Jia, and Luce (2000) for a discussion of additive error terms and other models in which the preference parameters are viewed as random variables.

For convenience of notation let $u^0(R)$ be the utility the subject would derive from obtaining one unit of her reference product and zero units of the other two products with the reference point r^0 she has prior to negotiating with agent A. Thus, for a subject in treatment 1, for example, $u^0(R) = u(\langle 1,0,0 \rangle \mid r^0) = m(\langle 1,0,0 \rangle) + n(\langle 1,0,0 \rangle \mid r^0) + \epsilon$. Let $u^0(N)$ be defined analogously for the non-reference product.

In the subject pool, utility derived from receiving either product 1 or product 2 is distributed according to some unobserved and unknown distribution function. However, as the usage of issues 1 and 2 as reference and non-reference product is balanced, the difference of utilities $u^0(R) - u^0(N)$ is a-priori distributed according to a distribution function F^0 that has mean zero and density f^0 symmetric around zero. A negative utility difference implies the choice of the non-reference issue and a positive difference the choice of the reference issue.

Ex-Post Utility Differences

In the rational choice model, there is no change of utility functions during the negotiation with agent A. Let the utility derived from the reference or the non-reference issue at the time when choosing among offers by agent C be labeled $u^C(R)$ and $u^C(N)$, respectively. The difference $u^C(R) - u^C(N)$ is unaffected by the negotiation with agent A, and hence, it is distributed according to F^0 as well.[10]

In the attachment effect model, on the other hand, utility is influenced by the shift of the reference point. The reference point is assumed to be higher on the reference issue than on the non-reference issue and, thus, loosing on the reference issue would be perceived stronger than loosing on the non-reference issue. Hence, $u^c(R)$ is relatively higher than $u^C(N)$. The distribution of the utility difference $u^C(R) - u^C(N)$ is affected by this shift in form of a shift to the right, i.e. to higher difference values (and potentially by a change of the shape of the distribution). Ex-post, the utility difference is distributed according to some function F^C with $F^C(d) < F^0(d)$ for any utility difference d—higher utility difference become likelier.

Uncertainty

The absolute value of the utility difference is assumed to be inversely proportional to preference uncertainty—a lower utility difference comes

---

[10] Note that the symmetry of f^0 around zero implies that the fraction of subjects preferring the reference product is equal to the fraction preferring the non-reference product. As the usage of products 1 and 2 as reference and non-reference product is interchanged between treatments 1 and 2 and there is an equal number of subjects in both treatments, this in turn implies $\pi_\succ^{T1} = \pi_\succ^{T2}$ in line with Section 5.1.5. However, the line of argumentation in Section 5.1.5 is more general as it does not assume a cardinal multi-issue utility function.

with a higher preference uncertainty. The intuition is that the closer one unit of product 1 is to one unit of product 2 in terms of utility, the more uncertain a subject is about which of the two to choose (or, to put it more technically: the higher the likelihood that the random error in the perception of utility determined the sign of the utility difference). Preference uncertainty is highest when the subject assumes to gain the same utility from products 1 and 2 (including the respective error term).

Decision-making is a mental procedure of information processing; it involves the sub-processes information acquisition, information evaluation, and expression of a decision (cf. Sec. 2.3.1). Preference uncertainty affect the information evaluation phase and the heuristic employed here. The task factors (i.e. the structural characteristics of the choice) are the same for all subjects; context factors might, however, differ. The context depends on the specific alternatives under consideration in a single decision task. Objectively they are the same for all subjects but they might be perceived differently depending on the prior negotiation. If so, the context might influence the choice of an information processing heuristic and this difference might be reflected in the data (cf. Sec. 2.3.1). Trade-offs across issues can, for example, be influenced by the choice of a specific heuristic (Nowlis and Simonson, 1997).

According to the rational choice model, utility differences are distributed with mean zero and density symmetric to zero. Thus, the expectation of the absolute value of the utility difference is the same for negative and positive utility differences. Let $E(\cdot)$ be the expected value and $abs(\cdot)$ the absolute value operator. Then $E(abs(u^C(R) - u^C(N)) \mid u^C(R) - u^C(N) < 0) = E(abs(u^C(R) - u^C(N)) \mid u^C(R) - u^C(N) > 0)$ holds. Thus, average preference uncertainty is the same for subjects with negative utility difference and positive utility difference and—with the utility difference implying the choice of the reference or the non-reference product—average preference uncertainty is the same for subjects choosing the reference product and subjects choosing the non-reference product.[11]

In the attachment effect model, on the contrary, the distribution F^C of utility differences is shifted to the right. Given random assignment of subjects to treatments, the density f^C is still symmetric around the mean, the mean however is greater than zero. Thus, the inequality $E(abs(u^C(R) - u^C(N)) \mid u^C(R) - u^C(N) < 0) <$

---

[11] It is assumed that indifferent subjects $(u^C(R) = u^C(N))$ choose either product with the same probability.

$E(abs(u^C(R) - u^C(N)) \mid u^C(R) - u^C(N) > 0)$ holds, i.e. the expected absolute utility difference is smaller for subjects showing a negative difference than for subjects showing a positive difference. Consequently, subjects choosing the non-reference product (negative utility difference) tend to have higher preference uncertainty than subjects choosing the reference product (positive utility difference).

Proxies for Preference Uncertainty

Preference uncertainty cannot be observed directly. Instead, response time of the choice among offers by agent C and reported subjective complexity of this choice are taken as proxies for preference uncertainty. Obviously, both measures—response time and subjective complexity—should be positively correlated if they are both proxies for one and the same underlying construct, namely preference uncertainty. However, several factors might confound their association: Firstly, complexity is just measured on a five-step scale whereas time is continuous and, secondly, subjects' ex-post perception of preference uncertainty might differ from their actual uncertainty during making the choice. The latter assumption stems from two psychological concepts: self-perception theory and dissonance theory. (See Curhan, Neale, and Ross (2004) for a recent discussion of these theories with respect to negotiations.)

Self-perception theory argues that someone who observes his own choice treats this as evidence for her own preferences just as she would treat the choice by someone else as revelation of his preferences (Bern, 1967, 1972). Thus, once the subject has chosen an offer by agent C, she infers that she prefers this offer and, as she prefers it, it cannot have been too difficult to choose it. Hence, the subject's ex-post judgment on how complex the choice was might differ from the actual complexity during the choice.

Furthermore, dissonance theory argues that choice creates cognitive dissonance as one has to forgo beneficial aspects of the unchosen alternative and has to accept unattractive features of the chosen alternative (Festinger, 1957; Festinger and Aronsons, 1960).[12] Humans do, however, according to Festinger dislike cognitive dissonance. To reduce dissonance, one tends to value the chosen alternative more positively and the unchosen alternative less positively after the choice. Hence, preference uncertainty diminishes and, again, the ex-post perception of complexity might differ from the true complexity during the choice.

---

[12] This argument holds for choices that involve multi-issue trade-offs—as the choice among offers by agent C—but not for single-issue choices or choices with a dominant option.

To summarize, response time and subjective complexity can both be employed as proxies for preference uncertainty and should corroborate one another. Response time is, however, the more reliable of the two as measurement is more subtle and less prone to confounding cognitive processes applied ex-post.

Response Time

Response time of a subject is the time that elapses from the moment at which a new screen is shown by the experiment software to the moment when the subject sends her answer to the server. Response times are measured separately for each subject and each screen shown like, e.g., the product selection, a new offer by agent A, the offer by agent B, the offer by agent C, and the overall result. Subjects were not informed that time would be measured and the measurements had no influence on the negotiations or the subjects' payoff. The response to the offers by agent C is of special interest here.

Response time and preference uncertainty are assumed to be positively correlated (Fischer, Luce, and Jia, 2000). The more uncertain a subject is in choosing product 1 or product 2, the longer she is likely to take for choosing either product. In several studies on choice tasks, response time is taken as proxy for effort devoted to that task (e.g. Hutchinson, Raman, and Mantrala, 1994; Haaijer, Kamakura, and Wedel, 2000). The reasoning is that a high utility difference between two alternatives makes the choice obvious and, thus, it requires low effort. Low effort in turn implies a short response time. Evidence for a decrease of response times with increasing difference in utility is presented by Tyebjee (1979), Böckenholt et al. (1991), and Bettman et al. (1993). The usage of response time as proxy for preference uncertainty follows from this line of argumentation as a high utility difference corresponds to a low preference uncertainty (Fisher, Luce, and Jia, 2000; Fisher, Jia, and Luce, 2000).

The different predictions for preference uncertainty directly transfer to response times. Let t^R be the time a subject takes for choosing her reference product and t^N the time for choosing the non-reference product. The rational choice model predicts no systematic difference ($t^R = t^N$) whereas the attachment effect model predicts lower preference uncertainty, and hence, lower response times for subjects choosing their reference product ($t^R < t^N$).

Subjective Complexity

At the lower end of the screen that displays the offers by agent C, each participant is asked to specify as how complex she perceives the choice among offers. Complexity is measured on a five-step Likert scale ranging from 'very easy' to 'very complex'. It is assumed that resolving a high preference uncertainty is perceived as complex whereas a subject that clearly knows which of the offers she prefers perceives this choice as easy. Whether this assumption is justified, can be assessed by looking at the same complexity measure for the choice of either the agreement with agent A or the offer from agent B—due to the dominance relation of these alternatives, this choice is assumed to be fairly easy.

Let c^R be the complexity a subject reports given that she chooses her reference product and c_N the complexity given she chooses her non-reference product $(c^R, c^N \in \{1, 2, 3, 4, 5\})$. Again, the predictions for preference uncertainty directly transfer to this proxy: The rational choice model predicts no systematic difference $(c^R = c^N)$ whereas the attachment effect model predicts lower subjective complexity for subjects choosing their reference product $(c^R < c^N)$.

5.2 Foundations of the Analysis

This section outlines the statistical tests to be used in the analysis and defines a distance measure in the offer space to analyze concessions made during the alternating offer negotiation.

5.2.1 Statistics to be Used

In order to analyze the experimental results, the statistical analysis in Section 5.3 applies several hypothesis tests: two-dimensional contingency tables are analyzed by means of the Fisher exact test and χ^2 tests on independence and homogeneity. Moreover, Wilcoxon rank sum tests are utilized to analyze the central tendency of observations with respect to a single dimension and correlation tests as well as binomial tests are performed.

Additionally, Section 5.4 employs a maximum likelihood estimation and likelihood ratio tests to test for significance of parameter estimates and construct confidence intervals. The foundations for this maximum likelihood analysis are not outlined here but discussed in Section 5.4.1 where they can be directly applied to the structure of the model and the data.

The software package R is used in its version 2.1.0 for computation of test statistics and error probabilities and for generating the figures (R Development Core Team, 2005). R provides standard implementations of all tests for the analysis of the data. In particular, the functions *fisher.test* (Fisher exact test), *chisq.test* (χ^2 test on homogeneity and χ^2 test on independence), *wilcox.test* (Wilcoxon rank sum test), *cor.test* (Correlation test), and *binom.test* (binomial test) are used. The respective parametrization of the test is described in more detail below.

Fisher Exact Test

The conditional Fisher exact test ('Fisher test' for short) is used to analyze the significance of association between two dimensions in 2×2 contingency tables (Fisher, 1934, 1935b).[13] The Fisher test is a common test for 2×2 contingency tables and discussed in many textbooks; see e.g. Agresti (1990, Ch. 3) for a formal and Sheskin (2004, Test 16c) for a more verbose discussion.

The (undirected) null hypothesis tested on a 2×2 contingency table is that in the underlying populations represented by the samples the proportion of observations in row 1 that falls in column 1 is equal to the proportion of observation sin row 2 that falls in column 1.

The assumptions underlying the test are as follows (Sheskin, 2004, Test 16c): (1) Categorial or nominal data (i.e. frequencies) for 2×2 mutually exclusive categories are used; (2) the data analyzed represents a random sample of independent observations (i.e. each subject contributes at most one observation); (3) the row and column marginal totals are fixed by the researcher prior to data collection. The first two assumptions are uncritical for the tables the test will be applied to in the following. The third assumption, on the other hand, requires further discussion as it is rarely met—the median test in the version that applies a Fisher test (instead of e.g. a χ^2 test) to the resulting contingency table is one of the rare exceptional cases in which both row and column marginals are fixed in advance (Sheskin, 2004, Test 16e).

Over decades there has been a lively debate among statisticians on the applicability of the *conditional* Fisher exact test. The argumentation against the test mainly is that it conditions inference on both margins where only one margin is fixed by most experiment designs and that the test is inherently conservative. Barnard (1945, 1947), for example, was one of the early opponents of this procedure and proposed an *unconditional* test for 2×2 contingency tables that only assumes

---

[13] The test was almost simultaneously proposed by Fisher, Yates (1934), and Irwin (1935). Hence, the test is sometimes denoted as Fisher–Irwin test.

the distribution of one margin as fixed. (Note however, that Barnard (1949, 1984) revoked his argumentation against Fisher's test later on.) See e.g. D'Agostino, Chase, and Belanger (1988) for a more recent argumentation against Fisher's conditional test.

In the analysis of the data gathered in this experiment, Fisher's exact conditional test is nonetheless applied to 2×2 contingency tables even if the assumption that frequencies are conditioned on both sets of marginal totals is not met. With this decision, the analysis follows the understanding of Yates (1984): firstly, assuming both marginals as fixed for inference eliminates nuisance parameters and, secondly, the margins contain little information on association anyways. This interpretation is corroborated by, for example, Cox (1984) and Barnard (1984) (both discussing Yates' article), by Little (1989) (discussing the article by D'Agostino, Chase, and Belanger), and by several textbooks on statistical procedures like Agresti (1990, Ch. 3) and Sheskin (2004, Test 16c).

The Fisher test is applied to 2×2 contingency tables either as one-sided test (directed null hypothesis either 'odds ratio is less than unity' or 'odds ratio is greater than unity') or as two-sided test (undirected null hypothesis 'odds ratio is equal to unity'). The odds ratio is a measure of association in contingency tables. A value of unity means that there is no association between the table's dimensions. Verbally, the undirected null hypothesis can be stated as hypothesis that in the underlying population the samples represent, the proportion of observations in row 1 that fall in column 1 is equal to the proportion of observations in row 2 that fall in column 1.

Results are reported by stating the number of observations (n), the conditional maximum likelihood estimate of the odds ratio, and the p-value. The estimation of odds ratio is thereby computed with the iterative procedure outlined by Cornfield (1956). Note, that this conditional maximum likelihood estimate proposed by Fisher (1935b) is slightly different from the unconditional maximum likelihood estimate that can directly be calculated from the observed cell frequencies.

The test as outlined above is readily available in R as *fisher.test*. A directed null hypothesis is obtained by the parameter *alternative="g"* or *alternative="l"*, respectively, and the undirected null hypothesis by *alternative="t"*. Default values are used for all other parameters.

χ^2 Test on Homogeneity

For contingency tables larger than 2×2, χ^2 tests are used. The χ^2 *test on homogeneity* is employed when two or more independent samples are categorized on a single dimension. The null hypothesis evaluated is that in the underlying population represented by the samples the categorization is homogenous for all samples. The assumption is that the number of observations for each sample is determined by the researcher, whereas the categorization along the measurement dimension is endogenous to the experiment (Sheskin, 2004, Test 16). The assignment of subjects to treatments is, for example, the partition of subjects in two samples. On the contrary, their choice of accepting offer 1 or offer 2 made by agent C is a self-categorization. Thus, a χ^2 test on homogeneity is used to test whether the choice of either offer is homogenous in both treatments. The test is inherently non-directional.

χ^2 Test on Independence

Another variant of the χ^2 test is the χ^2 *test of independence*. Both variants of the χ^2 test for contingency tables are computationally identical and just differ in the hypothesis evaluated. The test of independence is employed when a single sample is categorized on two dimensions. It evaluates the general hypothesis that the categorization in these two dimensions is independent of one another, i.e. there is a zero correlation between them (Sheskin, 2004, Test 16). The test is employed when neither the row nor the column sums of a contingency table are determined by the researcher. An example is the choice of a product from category 1 and a product from category 2 at the beginning of the experiment. All subjects constitute one sample and the question to evaluate is whether choices of the products are independent of one another.

The assumptions underlying both variants of the χ^2 test are firstly the usage of categorial or nominal data (i.e. frequencies) for $r \times c$ mutually exclusive categories and, secondly, that the data represents a random sample of independent observations (Sheskin, 2004, Test 16c). Generally, it is assumed that the test statistic follows a χ^2 distribution and p-values are computed accordingly. In case the expected frequency of one or more cells in the contingency table is low, the approximation by the χ^2 distribution might be inaccurate. In these cases, the p-value is computed by a Monte Carlo simulation with 10^6 replicates. There is no common agreement on what exactly a 'low' expected cell frequency is. In analogy to the suggestion by Sheskin (2004, Test 16), the simulation instead of approximation of the distribution is used for contingency tables with less than 40 observations or when at least one expected cell frequency is less than 5.

χ^2 tests are reported in form of the number of observations (n), the value of the test statistic (χ^2) using median centering, and the p-value. In case the p-value is obtained by a Monte Carlo simulation, this is indicated by 'sim. p-value'. Otherwise, i.e. when the χ^2 distribution is used for computing the p-value, the degrees of freedom (df) are given additionally.

All χ^2 tests are performed with a standard implementation in the R software package (*chisq.test*). The approximation by the χ^2 distribution as well as the Monte Carlo simulation are readily available in R. The simulation uses the algorithm presented by Patefield (1981) and is activated via the parameters *simulate.p.value=T* and *B=10^6*.[14] Default values are used for all other parameters.

Wilcoxon Rank Sum Test

The Wilcoxon rank sum test is used to test the central tendency of two independent samples (Wilcoxon, 1949). Oftentimes, it is also referred to as Mann–Whitney U-test (Mann and Whitney, 1947). The Wilcoxon rank sum test is a standard test employed with rank-order data. It is extensively discussed in most textbooks on inferential statistics like Hollander and Wolfe (1973, Ch. 4) and Sheskin (2004, Test 12).

The hypothesis evaluated is whether two independent samples represent two populations with different median values. To account for the rank-ordering of the data, the question can also be stated as: do two independent samples represent two populations with different distributions with respect to the rank-orderings of the scores in the two underlying population distributions? (Sheskin, 2004, Test 12)

Assumptions underlying the Wilcoxon rank sum test are: (1) Each sample is randomly selected from the population it represents, (2) the two samples are independent of each other, (3) the original variable observed (which is subsequently ranked) is a continuous variable,[15] and (4) the underlying distributions are of identical shape. (Sheskin (2004, Test 12) based on Conover (1999). See e.g. Seifert (2005, Ch. 3.8) for a discussion of the assumption on identically shaped distribution functions.)

For large samples, i.e. more than 50 observations in one sample, or ties, the distribution of the normalized test statistic is approximated by the normal distribution. Yates' correction for continuity is applied in these cases for not inflating the type I error rate by the approximation

---

[14] '10^6' is the notation used by the statistics software to denote '10^6'.

[15] This assumption is common to many rank-based tests. It is oftentimes not adhered to.

of a discrete distribution by a continuous one. For smaller samples, exact p-values are calculated.

Wilcoxon rank sum tests are reported by means of the number of observations in both samples (n_1 and n_2), the test statistic (W), and the p-value. If continuity correction and normal approximation are used, this is indicated. Computation of the test is performed with the function *wilcox.test* in the R software package using the parameter *paired=F*. A directed null hypothesis is obtained by the parameter *alternative="l"* or *alternative="g"* and the undirected null hypothesis by *alternative="t"*. Default values are used for all other parameters.

Correlation Test

Spearman's rank correlation coefficient is a bivariate measure of association between two variables (Spearman, 1904). It is employed with pairs of rank-order data. The population parameter is denoted ρ_S, the sample estimate r_S ($-1 \leq r_S \leq 1$). The absolute value of the correlation measure gives the strength of relationship.

The undirected null hypothesis tested is: In the underlying population represented by the sample, is there a significant monotonic relationship between the two variables? Or, to account for the rank-ordering, the hypothesis can be stated as: In the underlying population represented by the sample, is the rank correlation between subjects' scores on two variables some value other than zero ($H_0 : \rho_S = 0$)? (Sheskin, 2004, Test 29) The test relies on the assumptions that the sample is randomly selected from the population it represents, each element of the sample (i.e. each subject) contributes scores on two variables, and the data is in rank-order format.

The Pearson product-moment correlation coefficient is an alternative and commonly used bivariate measure of association (Pearson, 1896, 1900). It is not applied in the following analysis, as one assumption underlying the calculation of the Pearson product-moment correlation is that the two variables follow a bivariate normal-distribution. This can, however, not be assumed for the data two which a correlation test is applied in the following analysis. Thus, Spearman's rank correlation coefficient is used.

Test results are reported via the number of observations (n), the test statistic (S), and the p-value are reported. The computation follows the algorithm proposed by Best and Roberts (1975).

The computation of r_S and the corresponding test are readily available in R as *cor.test* using the parameter *method="spearman"*. Additionally, a directed null hypothesis is obtained by the parameter *alter-*

native="l" or *alternative="g"* and the undirected null hypothesis by *alternative="t"*. Default values are used for all other parameters.

Binomial Test

The final test employed in the non-parametric analysis of the experimental data is the binomial test. A series of n independent observations is randomly selected from a population and each observation can be classified in one of two mutually exclusive categories. The null hypothesis is as follows: In the underlying population represented by the sample, are the observed frequencies for the two categories different from their expected frequencies?

Computation is outlined by Sheskin (2004, Test 9). Results are reported by the number of observations (n), the number of so called successes, i.e. the number of observations that belong to the first of two categories, and the p-values.

The test is implemented in R as *binom.test*. The expected probability of success is specified by the parameter *p*, e.g. *p=0.5*. A directed null hypothesis is obtained by the parameter *alternative="l"* or *alternative="g"* and the undirected null hypothesis by *alternative="t"*. All other parameters are used with their default values.

5.2.2 Distances in the Agreement Space

A distance measure in the agreement space X can be used to assess the distance between two offers. Thus, it can be used to describe concessions made by the negotiators in the alternating offer exchange. The normalized Euclidean distance is employed. For two points $x^1 = \langle x_1^1, x_2^1, x_3^1 \rangle \in X$ and $x^2 = \langle x_1^2, x_2^2, x_3^2 \rangle \in X$ it is defined as

$$d_E(x^1, x^2) = \sqrt{\frac{1}{3}((\frac{1}{2}(x_1^1 - x_1^2))^2 + (\frac{1}{2}(x_2^1 - x_2^2))^2 + (\frac{1}{40}(x_3^1 - x_3^2))^2)}$$

Thus, it is a common Euclidean distance with two normalizations: Firstly, the maximum distance in each issue (i.e. dimension of the three-dimensional agreement space) is normalized to values between zero and unity (factors $\frac{1}{2}$ or $\frac{1}{40}$ for single-issue distances) and, secondly, the overall distance is normalized to the interval zero to unity (factor $\frac{1}{3}$ for the sum of squared distances).

The minimum distance between two offers is zero; it is obtained if and only if offers are identical ($x^1 = x^2$). Unity as the maximum distance possible in the agreement space is obtained only for the two offers $x^1 = \langle 0, 0, 0 \rangle$ and $x^2 = \langle 2, 2, 40 \rangle$. The distance from the origin to the midpoint of the agreement space is, for example, $d_E(\langle 0, 0, 0 \rangle, \langle 1, 1, 20 \rangle) = 0.5$.

Interpretation

The distance d_E is a distance, not a cardinal multi-issue utility function. Thus, it reflects the mathematical (dis)similarity of two offers but not necessarily their (dis)similarity as it is perceived by a subject. The distance d_E to the origin of the agreement space could be interpreted as multi-issue utility function. This would suppose that the respective decision-maker values the maximum level on each issue individually equally strong. In other words, obtaining 40 units of product 3 and zero units of the other two products would yield the same utility as 2 units of product one and zero units of the other two products, for example. In the absence of knowledge about the subjects' utility functions, this is a crude approximation.

Alternative Measures

Alternative distance measures can be employed: Firstly, a subject's utility function could be used to measure her perceived difference between to offers. However, the subjects' utility functions are not known and not elicited do to reasons outlined in Section 5.1.5.

Secondly, the retail price of an offer could be used as approximation of its utility to subjects. However, if retail price is a good proxy for utility, then not a single subject should choose to negotiate over either key cords or coffee mugs as these two goods have lower retail prices than the other goods in their respective category.

Thirdly, and finally, a dominance relation among points in the agreement space could be used as introduced by Vetschera (2004a) and used by Block et al. (2006). The dominance relation does not assume trade-offs across issues. However, the usage in this context has two problems: (1) the third product would play a highly prominent role in comparing two offers, as the domain of possible values ranges from zero to forty instead of just from zero to two. Thus, the relatively low retail price of a single unit of product 3 would not be reflected. (2) The measure is as well only a crude approximation of utility and cannot resolve indifference among several quite distinct offers.

To summarize, in the absence of knowledge about individual utility functions, any measure that compares different points in the agreement space can just be an approximation of how subjects perceive the comparison of offers. Which of several approximations is best cannot be decided unanimously. In the following analysis, the normalized Euclidean distance d_E is employed. As the interrelation of this measure and subjects' utility cannot be determined precisely, d_E is solely used for characterizing the alternating offer exchange between subjects and agent A. The distance is, however, not used for the hypotheses on treatment effects outlined in Section 5.1.5.

5.2.3 Overview of the Analysis

The following Table 5.1 gives an overview on the statistical analysis of the data gathered in the lab experiment. Descriptive statistics are used throughout the entire analysis and, thus, they are not specifically mentioned here.

Table 5.1. Inferential statistics used in the data analysis

Section	Inferential statistics	Page
5.3 Non-parametric analysis		179
5.3.1 Overview of the data	Fisher test Binomial test Correlation test χ^2 test on homogeneity	179
5.3.2 Binary choice	Fisher test χ^2 test on independence χ^2 test on homogeneity	190
5.3.3 Sources of utility	χ^2 test on independence χ^2 test on homogeneity	195
5.3.4 Preference uncertainty	Correlation test Wilcoxon rank sum test	200
5.4 Parametric analysis		206
5.4.2 Parameter estimation	Maximum likelihood estimation Likelihood ratio test	211
5.4.3 Reliability of the estimation	Wilcoxon rank sum test Correlation test	215

5.3 Non-Parametric Analysis

This section tests for treatment effects based on three distinct measures: binary choice of the preferred product (Sec. 5.3.2), sources of utility (Sec. 5.3.3), and preference uncertainty (Sec. 5.3.4). Theoretical predictions were derived from the rational choice model and the attachment effect model in Section 5.1.5. Prior to this tests, an overview of the data is giving in the following.

For hypothesis testing, a 5% significance level is used throughout the analysis. The term *negotiator* is used when there is no need to differentiate among subjects and the agent and the symbol '#' in tables stands for 'number of'.

5.3.1 Overview of the Data

This section presents an overview on the subjects that participated and their behavior in the alternating offer exchange with agent A. The analysis starts with the subjects' gender, age, course of studies, and experience. Out of 82 participants, 67 are male and 14 are female.[16] One subject did not answer this optional question. The median of subjects' age is 23. Table 5.2 displays the respective distribution, the median is highlighted by a surrounding box.

Table 5.2. Subjects' age in years

Age in years	≤ 20	21	22	23	24	25	26	27	≥ 28
#subjects	6	18	12	13	12	10	4	2	5

The subjects' courses of studies are displayed in Figure 5.7. Most subjects (46 out of 82) study Economics or related courses (Business Engineering or Information Engineering and Management). The subject pool is common for experiments at the University of Karlsruhe.

The participants' experience with laboratory experiments is shown in Table 5.3. Most subjects had already participated in lab experiments before; the median value is 3 experiments. Out of these, most subjects did not yet participate in more than 1 experiment on negotiations.

---

[16] The small proportion of female subjects is not uncommon, as just about 25% of students at the (Technical) University of Karlsruhe are female.

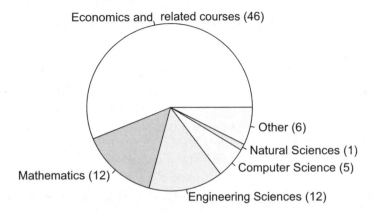

Fig. 5.7. Subjects' courses of studies

Table 5.3. Subjects' experience with experiments in general and with negotiation experiments

	#experiments					
	0	1–2	3–4	5–6	≥ 7	no answer
#subjects (all lab experiments)	8	26	$\boxed{21}$	13	13	1
#subjects (thereof negotiations)	31	$\boxed{31}$	12	3	2	3

Product Selection

Participants have the choice over which specific products they want to negotiate: For product 1, they can choose among a key cords and an espresso pots and for product 2, they can choose either a coffee mugs or a thermos flasks. Both choices are entered on the same screen but the selection of a product from category 1 and a product from category 2 are conceptually independent of one another. Table 5.4 summarizes the product combinations selected by subjects.

The data comes, as before, from 82 subjects: 41 in treatment 1 and 41 in treatment 2. Overall, 84 subjects participated in the experiment. However, the design bases on paired observations from subjects who choose the same products in the beginning and are assigned to corresponding negotiation strategies by agent A. This design allows for the possibility that for each product combination one observation might have to be dismissed in case there is an even number of subjects that choose this combination. In fact, this happened for two combinations (key cord / coffee mug and espresso pot / thermos flask). Thus, two

Table 5.4. Results of product selection

		product 2		
		coffee mugs	thermos flasks	Σ
product 2	key cords	6	16	22
	espresso pots	16	44	60
	Σ	22	60	82

observations are dismissed. These stem from two subjects that chose the specific combinations in the last session of the experiment. The decision to dismiss the data of exactly these two subjects was made automatically by the experiment software prior to and independent of the subjects' behavior in the negotiations and subsequent choices. The entire data of the two subjects is reported in an appendix (which is available upon request) along with the data from all other subjects. For the remainder of the analysis, however, these additional two observations are dismissed.

A Fisher exact test is applied to the Table 5.4 with the null hypothesis that in the underlying population represented by the sample the proportion of subjects that choose to negotiate over coffee mugs is independent of whether the respective subject chooses to negotiate over key cords or espresso pots. The null hypothesis can not be rejected (n = 82, odds ratio = 1.031, p-value > 0.5). Thus, the choice of product 1 appears to be independent of the choice of product 2 not only on a conceptual level but as well reflected in the subjects' choices. For both product categories, the second alternative, i.e. the espresso pots and the thermos flasks, is chosen by 60 out of 82 subjects. This choice is significantly non-random for both product categories individually, as a binomial test on the null hypothesis of equal probability choice shows (n = 82, 60 successes, p-value < 0.001).

RESULT 5.1: *Subjects choose products 1 and 2 independently of one another and in both categories one of the products is chosen more frequently than the other.*

The fact that the choice of products is independent of one another strengthens the assumption of monotone preferences for each issue individually. This assumption was made in Section 5.1.5 to derive the hypothesis on consumption and gain/loss utility.

Final Payment

At the end, the contracts of 35 subjects were based on their respective agreement with agent A or agent B and the remaining 47 subjects got the contract with agent C. On average each subject received 0.61 units of product 1, 0.82 units of product 2, and 5 units of product 3. In addition, the fix amount of € 5 cash was paid to each participant. Overall, this amounts to 9 key cords, 41 espresso pots, 18 coffee mugs, 49 thermos flasks, 411 CDs, and € 410 cash.

Alternating Offer Negotiation

The purpose of this section is twofold: Firstly, it gives an overview on the subject's behavior in the alternating offer exchange with agent A. The subjects' behavior is not tested against theoretical predictions; the presentation rather outlines that the behavior is in line with a common sense understanding of a 'normal negotiation', i.e. an repeated exchange of offers with several gradual concessions and trade-offs. The second purpose of the section is to compare subject behavior with agent behavior. Agent behavior is not too different from subject behavior. Again, this corroborates the assumption that the alternating offer negotiation is not a specifically uncommon interaction of the negotiators but rather a 'normal negotiation'. Thus, this section is important for determining in how far a generalization of results to other negotiations is valid.

Comparing behavior to a game theoretic equilibrium model and having equilibrium strategies for agent A would be preferable. However, there is no such model for the bilateral multi-issue alternating offer game under incomplete information studied here.

Each negotiation starts with an initial offer by agent A. Subsequently, offers are exchanged in turns. Thus, an overall even number of offers in a negotiation indicates that agent A accepted an offer made by the subject. An odd number of offers, on the contrary, indicates that the subject accepted one of agent A's offers. The only exceptions are negotiations with exactly twelve offers. When this limit is reached, the negotiation might either end with the agent accepting the subject's last offer or with the arbitrator stepping in and imposing an agreement.

In 33 negotiations, the agent accepted the respective subject's offer, in 25 negotiations, the subject accepted the agent's offer, and in the remaining 24 negotiations the arbitrator determined the agreement. The limit of twelve offers is reached by 34 subjects. In 10 of these negotiations the subject's last offer is accepted by the agent. In the remaining 24 cases the arbitrator imposed the agreement, as mentioned before. Figure 5.8 displays the number of offers per negotiation in a histogram. The 34 negotiations with twelve offers each are split up in the arbitrated cases and the ones in which an agreement was reached with the last offer.

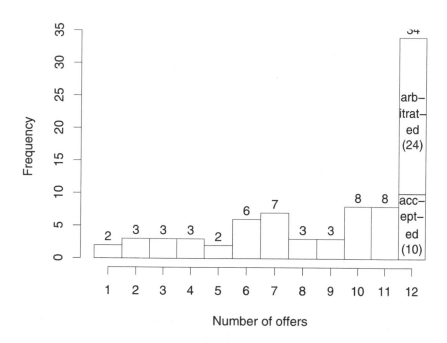

Fig. 5.8. Number of offers in negotiations with agent A

Negotiation Tactics

As soon as a negotiator makes two offers, the transition of offers can be observed. It can be measured how two subsequent offers differ from each other. The following four categories of so called *negotiation tactics* are considered:

Similarity: The offer exactly resembles the negotiator's previous offer.
$(\forall\ k \in \{1,2,3\} : x_k^t = x_k^{t-2})$ [17]

Strict concession: From the negotiator's point of view, the offer is strictly dominated by her previous offer. The negotiator makes a concession to the counterparty on at least one issue without claiming more on any other issue.
$(\forall\ k \in \{1,2,3\} : x_k^t \leq x_k^{t-2} \text{ and } \exists\ k \in \{1,2,3\} : x_k^t < x_k^{t-2})$

Strict step back: Analogous to the strict concession, but to the advantage of the negotiator (at least it appears so in the short run). She claims more on at least one issue without giving the counterparty an advantage on any other issue.
$(\forall\ k \in \{1,2,3\} : x_k^t \geq x_k^{t-2} \text{ and } \exists\ k \in \{1,2,3\} : x_k^t > x_k^t)$

Trade-off: The negotiator makes a concession on at least one issue and in turn claims more on at least one other issue.
$(\exists\ k,l \in \{1,2,3\}, k \neq l : x_k^t < x_k^{t-2} \wedge x_l^t > x_l^{t-2})$

These four tactics are exhaustive given that a negotiator makes an offer that is not her first one. The terms *tactic* and *strategy* are used differently by several authors and especially by authors from different disciplines. Generally, strategies are long term plans of action to achieve a specific goal whereas tactics are immediate actions employed. The aforementioned characterizations of transitions from one offer to the next are called tactics here for two reasons: Firstly they are immediate actions at one point in time and a negotiator can employ several tactics to implement a general strategy. Secondly, the term tactic is used to clearly draw the distinction to the game theoretic notion of a strategy.

All four tactics outlined above are employed by subjects and are used by agent A in different offer sequences. Table 5.5 summarizes how often the tactics or combinations thereof are used. The table reads as follows: 1 subject employed similarity as only tactic (upper left corner), whereas not a single offer sequence by agent A included this as only tactic (upper right corner). Another 17 subjects chose strict concessions as only tactic. The agent used 6 corresponding offer sequences. Several '×' in a line indicate that a negotiator used several distinct tactics: 9 subjects used, for example, a combination of similarity and strict concessions without using the other two tactics. Note that the line without any '×' says that 8 subjects did not use any of the tactics and 5 times the agent did not use any. This is due to the fact that they did

---

[17] The notation is like in the previous sections: x_k^t stands for an offer of x on issue k at time t. It is assumed that the negotiator prefers higher values to lower ones on each single issue.

not make at least two offers, a necessary condition for identifying the tactics.

The lines labeled *#subjects* and *#offers* in the first column show how many different subjects utilize a tactic at least once and for how many offers they do so all in all. Strict concession, for example, is used by 67 out of the 82 subjects for overall 155 offers. Finally, the last two lines analogously represent data on the usage of tactics by agent A.

Table 5.5. Usage of negotiation tactics

#subjects	simi-larity	con-cession	step back	trade-off	# offer sequences
1	×				0
17		×			6
0			×		0
2				×	6
9	×	×			4
0	×		×		0
2	×			×	5
4		×	×		2
22		×		×	19
2			×	×	0
4	×	×	×		0
4	×	×		×	29
0	×		×	×	0
5		×	×	×	6
8					5
2	×	×	×	×	0
#subjects	22	67	17	39	
#offers	37	155	25	66	
	38	66	8	65	#offer seq.
	52	132	8	114	#offers

For the agent and subjects alike, the ranking of the usage of single tactics is the same (in terms of negotiators as well as in terms of offers): strict concessions are used most frequently, trade-offs are the second most popular tactic, similarity the third, and no other tactic is used as seldom as strict steps back. Furthermore, the rank order in that the agent uses the different patterns of tactics is positively correlated with the subjects' rank order of usage; the correlation is significantly different than zero (Spearman's rank correlation, $H_0 : \rho_S = 0$, n = 16, $r_s = 0.694$, p-value = 0.004).

RESULT 5.2: *In the alternating offer negotiation, subjects use different negotiation tactics and patterns thereof with approximately the same frequency as agent A does.*

The similar usage of tactics might not be by chance but could arise from subjects mimicking the agent's tactic. The interrelation can, however, not be the other way round as the agent's behavior is fixed in advance. Several studies support the assumption that negotiators create mutual understandings of rules of acceptable behavior; see e.g. Bazerman et al. (2000) for a review of studies on shared mental models. Table 5.6 shows how frequently a subject answers with either of the tactics, given that the agent used a specific tactic for the previous offer. Subjects employed, for example, the similarity tactic as response to the agent using similarity 4 times. For another 34 offers, subjects used strict concessions when the agent had previously used similarity, etc. Column sums correspond to the usage of tactics by subjects given in Table 5.5. Row sums, on the other hand, are slightly less than Table 5.5 would suggest, as offers by agents just contribute to Table 5.6 if they are followed by a counteroffer from the respective subject.

Table 5.6. Frequency of successions of negotiation tactics

		subject tactic, time $t+1$				
		simi-larity	con-cession	step back	trade-off	$\sum$
agent tactic, time t	similarity	4	34	1	11	50
	concession	17	67	17	25	126
	step back	2	3	0	3	8
	trade-off	14	51	7	27	99
	$\sum$	37	155	25	66	283

A χ^2 test on homogeneity is applied to Table 5.6. The null hypothesis of homogeneity of the usage of tactics by subjects following the usage of a tactic by the agent cannot be rejected (n = 283, $\chi^2 = 13.079$, sim. p-value $= 0.155$). Thus, there is no evidence that subjects mimic agent tactics.

Concessions

How strongly a negotiator makes concessions and moves towards the counterparty's ideal outcome can be measured via the normalized Euclidean distance d_E of her offers towards this ideal outcome (in the following the term *distance* is used as short version of 'distance from the counterparty's ideal outcome'). The definition of d_E was given in Section 5.2.2 along with a discussion of its interpretation as approximation of subjects' utility.

Figure 5.9a displays boxplots for the distances of offers. In the first plot, the data comes from all negotiators. As negotiations end after different number of offers (cf. Figure 5.8), the number of data points in each box decreases monotonically. This is indicated at the lower end of the plot as *number of negotiators*. Grey boxes represent the offers by agent A (odd offer numbers), white boxes the offers by subjects (even offer numbers).[18]

For Figure 5.9b, the data comes from the negotiations in which at most 6 offers were exchanged and Figure 5.9c displays the negotiations with exactly 12 offers. Hence, the second and the third plot summarize different (non-exhaustive) subsets of the data used for the first plot. The rational for these additional plots is that trends and differences that seem obvious in the first plot might be due to self selection of subjects continuing the negotiation or ending it.

With increasing offer number, the boxes representing subject offers and agent offers gradually 'move away' from each other. Note that this does not indicate that the negotiators' offers become less similar. The reason is that the distance for agents and subjects is measured to different points, namely the respective ideal outcome for the counterparty. Thus, the decrease for both groups shows that both parties in a negotiation gradually move towards the counterparty.

All three plots in Figure 5.9 suggest (at least) two observations: (1) Subjects tend to have a greater variation in their offers than the agent has, and (2) the distance tends to decrease from offer to offer for subjects and the agent alike. The difference in variation means that

---

[18] As usual for boxplots, a box itself ranges from the first to the third quartile, i.e. it covers half of the data points. The median is plotted in its inner body and the whiskers represent data points in 1.5 interquartile-ranges to both directions. Outliers are displayed as single points.

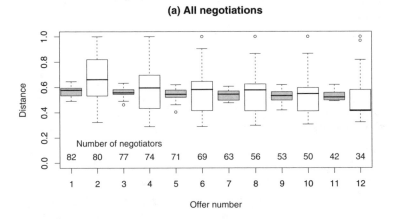

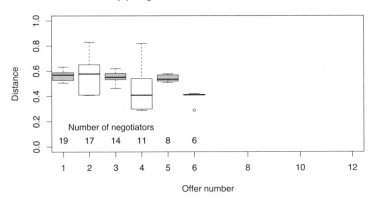

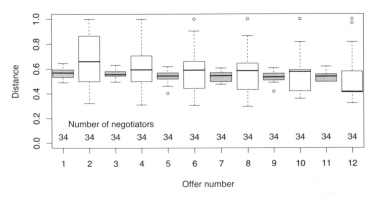

Fig. 5.9. Euclidean distance of offers to the counterparty's ideal outcome.

offers by different subjects are less similar (in terms of d_E) than offers from different offer sequences.

Decrease in Distances

Taken all negotiations together, there is a negative correlation of distances and offer numbers for the agent and subjects alike. The values of Spearman's rank correlation r_S are given in Table 5.7. As the offer number is ordinal, Spearman's rank correlation coefficient is used to analyze the association of distance and offer number. The same holds for the two subsets of the data, i.e. for negotiations with at most 6 offers and negotiations with exactly 12 offers.[19]

Table 5.7. Results of tests for correlation between distances of offers and offer numbers

	r_S	#offers	test statistic S	p-value
	All negotiations			
agent offers	-0.376	388	13393834	< 0.001
subject offers	-0.306	363	10409033	< 0.001
	$\leq$ 6 offers			
agent offers	-0.185	41	13604	0.250
subject offers	-0.441	34	9431	0.010
	12 offers			
agent offers	-0.327	204	1878069	< 0.001
subject offers	-0.347	204	1905322	< 0.001

A two-sided correlation test with the null hypothesis of zero correlation is applied to each of the six sets of offers individually (all

---

[19] Note that when interpreting the correlation coefficients in Table 5.7, a common mistake would be the following reasoning: 'Looking at all negotiations, the absolute value of r_S is bigger for agents than for subjects. Thus, the distance of agent offers reduces faster than the distance of subject offers: Agents make stronger concessions.' However, this conclusion is incorrect, as it ignores the different variance of distances depending on the nature of the negotiators.

To reach a valid conclusion on the strength of concessions, a regression analysis might be applied. However, this is omitted here for three reasons: (1) the measurement of distances is just a crude approximation of how subjects might perceive offers (cf. Sec. 5.2.2), (2) the necessary error term in a regression model cannot be assumed to follow a symmetric distribution for subject offers as the boxplots in Figure 5.9 show, and (3) the strength of concessions is not the focus of the analysis.

negotiations and two subsets; subject offers and agent offers for each). The results are as well displayed in Table 5.7. For the agent's and the subjects' offers alike, the correlation is significantly different from zero if one looks at either all negotiations, or negotiations with exactly 12 offers. For the subset of negotiations with at most 6 offers, the correlation is just significant for the subjects' offers but not for the agent's offers. Due to the small sample size, this is not surprising.

RESULT 5.3: *Subjects and the agent alike tend to make concessions and gradually approach the counterparty's ideal outcome with increasing offer numbers.*

Choice of a Dominant Alternative

The offer by agent B is constant for all subjects; it is $\langle 1, 1, 12 \rangle$. By design, this offer weakly dominates the agreement any subject has with agent A. The purpose of the design is that subjects accept the offer by agent B and all subjects have the same contract. 80 out of the 82 subjects accepted the offer as expected, 2 subjects—number 12 in treatment 1 and number 52 in treatment 2—did not.

The two subjects who rejected the offer are about average with respect to the time of consideration and the reported complexity of choice. Exceptional is that they had just 1 or 2 offers in the negotiation with agent A. Subject number 12 got an agreement of $\langle 1, 1, 0 \rangle$ and reported to be 'very satisfied' with obtaining zero CDs as well as with the overall agreement. Subject number 52 got an agreement of $\langle 1, 1, 3 \rangle$ and reported to be satisfied with the three CDs. Thus, one can speculate that these two subjects did not care about issue 3. Thus, they ended the negotiation quickly and did not bother to accept B's offer. As the two subjects do not appear to obtain positive utility from recordable CDs, it is assumed that there are no substantial wealth effects between these two subjects and the 80 other subjects that accepted the offer of $\langle 1, 1, 12 \rangle$. Furthermore, the two subjects belong to different treatments. Thus, in the following test for treatment effects, the question whether or not a subject accepted the offer by agent B is not discussed further.

5.3.2 Binary Choice

the simplest and clearest measure for a treatment effect is the subjects' decision in a binary choice among the two offers by agent C. The first offer is $\langle 1, 0, 0 \rangle$ for all subjects, the second is $\langle 0, 1, 0 \rangle$. Thus, choosing the first offer indicates a preference for the first product instead of the second and choosing the second offer indicates the reverse preference. In the following the terms 'choosing an offer' and 'choosing a product' will be used interchangeably, as they denote the same in this choice task. In case a subject is indifferent between the two offers, she has to resolve the ambiguity herself and choose either one.

Under the assumption of random assignment of subjects to treatments, the subjects' preferences in both treatments should not differ systematically, except if there is a treatment effect. The rational choice model predicts no systematic difference ($\pi_\succ^{T1} = \pi_\succ^{T2}$) whereas the attachment effect model predicts that subjects tend to choose the product for which their reference point is higher ($\pi_\succ^{T1} > \pi_\succ^{T2}$; cf. Sec. 5.1.5). Table 5.8 summarizes the subjects' choices.

Table 5.8. Contingency table on the effect of treatments on product choice

	offer 1 (product 1)	offer 2 (product 2)	$\sum$
treatment 1	26	15	41
treatment 2	12	29	41
$\sum$	38	44	82

From Table 5.8 it can be seen that subjects in treatment 1 are more likely to choose product 1 ($p_\succ^{T1} = 63\%$) and subjects in treatment 2 are more likely to choose product 2 ($p_\succ^{T2} = 29\%$)—just as predicted by the attachment effect model and in contradiction to the rational choice model. A directed Fisher exact test can reject the null hypothesis that in the underlying population represented by the subjects, the proportion of subjects in treatment 1 that select offer 1 is no greater than the proportion of subjects in treatment 2 that select offer 1 ($H_0 : \pi_\succ^{T1} \leq \pi_\succ^{T2}$, n = 82, odds ratio = 4.110, p-value = 0.002).

RESULT 5.4: *Offers in the alternating offer negotiation have a significant systematic influence on the subjects' preferences. As predicted by the attachment effect model, subjects tend to choose the product that was offered more frequently.*

Duration of Negotiations

The attachment effect model argues that reference points are influenced by offers. Thus, if all offers favor the same reference point, the subject should be more likely to adapt this reference point the longer the sequence of offers is. Hence, it can be assumed that a longer negotiation with agent A correlates with the likelihood of choosing the reference product. This interrelation is displayed by the histogram in Figure 5.10. One clearly sees that subjects with a high number of offers, e.g. 12 offers, tend to choose the reference product.

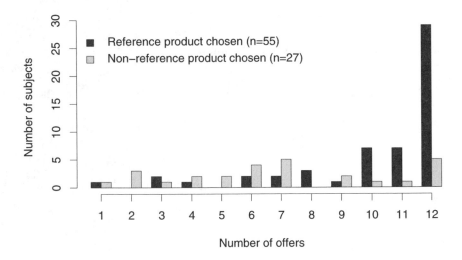

Fig. 5.10. Interrelation of the number of offers with agent A and the choice of the reference product

A χ^2 test on independence is applied to the 2×12 contingency table corresponding to Figure 5.10, i.e. the two dimensions are the choice of the reference or non-reference product and the number of offers. The null hypothesis of independence of these two dimensions in the underlying population represented by the sample can be rejected ($n = 82$, $\chi^2 = 30.940$, sim. p-value < 0.001).

RESULT 5.5: *The length of a negotiation with agent A has an impact on choosing the reference product or not. Subjects with a high number of offers tend to choose the reference product.*

Impact of the Arbitrator

Figure 5.10 suggests that the likelihood of choosing the reference product is especially high for subjects with exactly 12 offers. Some of these subjects reached an agreement as the software agent accepted their last offer, others because the arbitrator imposed an agreement. Thus, this exceptional peek in the histogram might be due to the arbitrator instead of the mere number of offers. However, it is not.

The frequency of arbitrator-imposed agreements and the choice of the reference product are shown in Table 5.9. For 10 subjects the agreement is their respective 12th offer made and for 24 subjects the agreement is imposed by the arbitrator (cf. Figure 5.8).

Table 5.9. Contingency table on the effect of the arbitrator on product choice

	Choice of reference product	non-reference product	$\sum$
Offer accepted	8	2	10
Arbitrator	21	3	24
$\sum$	29	5	34

An undirected Fisher exact test cannot reject the null hypothesis that in the underlying population represented by the sample, for subjects that end the negotiation with an accepted offer the proportion of subjects choosing the reference product is equal to the proportion of subjects choosing the reference product given that their agreement is imposed by the arbitrator (n = 34, odds ratio = 1.719, p-value > 0.5). Thus, the high frequency of subjects choosing the reference product given that they exchanged 12 offers with agent A is due to the mere number of offers and not to the arbitrator.

Possible Confounding Variables

Table 5.8 shows a clear treatment effect as predicted by the attachment effect model. However the following factors might have an influence:

- gender,
- age,
- course of studies,
- experience with experiments,
- experience with negotiation experiments,
- choice of two specific products at the beginning of the experiment, and
- session.

These factors are considered here as they might influence subjects' choices. However, summarizing the statistical analysis, one can say that for none of these factors a significant influence on choice can be shown. To exemplify this, the factor which comes closest to significance is outlined here; for the rest of the aforementioned factors, the statistics are given in an appendix which is available upon request.

Gender Effect

Table 5.10 shows how frequently female and male subjects choose the reference product as predicted by the attachment effect model (i.e. they choose product 1 in treatment 1 or product 2 in treatment 2) and how often they choose the non-reference product. The rational choice model predicts no systematic difference between reference and non-reference product. On the contrary, the attachment effect model suggests that subjects tend to choose their respective reference product.

Note, that the table only shows data from 81 subjects, as one subject did not answer the question for his or her gender (cf. Sec. 5.3.1). Thus, the 27 subjects choosing the non-reference product are the same as $15 + 12$ subjects choosing product 2 in treatment 1 or product 1 in treatment 2 as displayed in Table 5.8.

Table 5.10. Contingency table on the effect of gender on product choice

	Choice of		
	reference product	non-reference product	$\sum$
Female	6	8	14
Male	48	19	67
$\sum$	54	27	81

72% of male subjects but only 43% of female subjects choose the reference product. Thus, the data suggests that the behavior of male subjects is more likely to be in line with the behavioral prediction than the behavior of female subjects. Or, to put it differently, agent A seems to have less influence on the female subjects' preferences than on the male subjects' preferences. The difference is, however, just not significant at a 5% level (two-sided Fisher exact test on the null hypothesis that in the underlying population represented by the sample the proportion of female subjects choosing the reference product is equal to the proportion of male subjects choosing the reference product, n = 81, odds ratio = 3.312, p-value = 0.059).[20]

RESULT 5.6: *The proportion of male subjects choosing the reference product is higher than the proportion of female subjects choosing the reference product. The difference is, however, not significant.*

RESULT 5.7: *The preference for the reference product cannot be explained by a significant influence of gender, age, course of studies, experience with experiments, experience with negotiation experiments, choice of two specific products at the beginning of the experiment, or different sessions.*

5.3.3 Sources of Utility

Common microeconomic assumptions are consumption utility and monotonous preferences: if consuming a good gives positive utility, then having more of that good gives higher utility. Of course, there are many examples where consumption is beneficial initially but exaggerated consumption reduces utility—several pharmaceuticals can be taken as example—, but for the products used in the experiment and the small number of units, non-satiation is assumed.

The assumption of consumption utility is shared by the rational choice model and the attachment effect model. However, the attachment effect model adds a second 'source of utility', namely gain/loss utility. If an outcome is perceived as loss relative to the reference point this gives negative utility and if it is perceived as gain it gives positive utility. In rational choice, there is no gain/loss utility. Thus, evidence for gain/loss utility in the data would favor the attachment effect model over the rational choice model.

---

[20] The fact that data from one subject is missing here, might influence the test result. Given that one knows that the subject chose as predicted, the assumption that the subject is male would strengthen evidence for the gender effect. Moreover, given the base-rate of male and female subjects, this assumption is plausible. But even with this assumption, the gender effect is not significant (two-sided Fisher exact test on the same null hypothesis as before, n = 82, odds ratio = 3.380, p-value = 0.058.)

The Fisher test conditions inference on fixed marginals in both dimensions. This assumption is rarely met (cf. Sec. 5.2.1). As the result here is close to the significance level of 5%, an unconditional χ^2 test is reported as well. It supports the result of the Fisher test: χ^2 tests on independence on the null hypothesis that gender and choice of the reference or the non-reference product are independent of one another, n = 81, $\chi^2 = 4.318$, sim. p-value = 0.059 ('worst case' assignment of the missing data point (n = 82, $\chi^2 = 4.483$, sim. p-value = 0.058).

Satisfaction Ratings

At the end of negotiating with agent A—and after the potential arbitration—subjects are asked to specify how satisfied they are with the result for each single issue and with the overall agreement. Note, that there is no salient reward for this question and, thus, it has to be assumed that subjects are intrinsically motivated to report their satisfaction truthfully. Four possible reasons can be brought forward why subjects should be truthful: (1) entering the first (truthful) idea that comes to the mind of a subject is cognitively less demanding than reasoning about misrepresentation; (2) subjects' participate voluntarily, thus, they like the invitation the experimenter made and might reciprocate this favor by doing what they are asked to do; (3) subjects might appreciate being asked for their opinion; (4) finally, the answer has no effect on their reward for the experiment, thus, there is no reason not to be truthful. Anyways, even if (some) subjects are dishonest in reporting their satisfaction, there is no reason why this dishonesty should systematically relate to a treatment effect.

Satisfaction is specified on a five step Likert scale from 'very unsatisfied' to 'very satisfied'. These satisfaction values are taken as proxy for utility as a positive correlation of satisfaction and utility is assumed.

Besides consumption and gain/loss utility, other factors might influence subjects' satisfaction. Especially the length of the previous negotiation with agent A and whether it ended by arbitration or not could impact the ratings. However, it turns out that neither factor has a significant influence on the subjects' reported satisfaction. The detailed analysis is presented in an appendix which is available upon request. Thus, it seems justified to use satisfaction as proxy for consumption and gain/loss utility.

Consumption Utility

Consumption utility implies that a higher value for an issue should tend to imply higher satisfaction ratings by subjects ($k \in \{1, 2\} : s_k(0) < s_k(1)$; cf. Sec. 5.1.5). Table 5.11 shows the satisfaction of subjects in both treatments given an agreement of either 0 or 1 on issue 1 and Table 5.12 shows the corresponding data for issue 2. Medians are highlighted by boxes.

For issue 1, the median satisfaction is 3 if the subjects' agreement has a value of zero for this issue and the median is 4 if the subject negotiated to get one unit of this issue. For issue 2, the trend is the same, i.e. subjects getting one unit are more satisfied than subjects getting zero units. Thus, the inequality $s_k(0) < s_k(1)$ implied by the rational choice and the attachment effect model seems to hold in a between subject comparison.

χ^2 tests on independence are applied to both tables separately. The null hypothesis that agreement and satisfaction are independent in the underlying population represented by the sample cannot be rejected for issue 1 ($n = 82$, $\chi^2 = 6.141$, sim. p-value $= 0.187$) but it can be rejected for issue 2 ($n = 82$, $\chi^2 = 33.959$, sim. p-value < 0.001).

RESULT 5.8: *issues 1 and 2 favor the assumption that subjects satisfaction reflects consumption utility—for issue 2 this is significant, for issue 1 it is not.*

Table 5.11. Contingency table on consumption utility for issue 1

	issue 1		
	$s_1(0)$	$s_1(1)$	$\sum$
very dissatisfied (1)	4	5	9
(2)	3	13	16
(3)	3	13	16
(4)	1	16	17
very satisfied (5)	7	17	24
$\sum$	18	64	82

Table 5.12. Contingency table on consumption utility for issue 2

	issue 2		
	$s_2(0)$	$s_2(1)$	$\sum$
very dissatisfied (1)	5	3	8
(2)	1	13	14
(3)	1	18	19
(4)	0	23	23
very satisfied (5)	0	18	18
$\sum$	7	75	82

Gain/Loss Utility

Given a fixed outcome of either zero or one on an issue, gain/loss utility implies that a higher expected reference point comes with a lower
satisfaction ($k \in \{1,2\}, a_k \in \{0,1\} : s_k^R(a_k) < s_k^N(a_k)$, cf. Sec. 5.1.5).
Thus, given an agreement of one unit, for example, satisfaction should
be higher for the non-reference product than for the reference product. Loosely speaking, satisfaction is higher for an unexpectedly good
outcome than for a good outcome that was expected anyways. To guarantee independence of observations, issues 1 and 2 and agreements of
zero or one unit are analyzed separately in the following.

Table 5.13 shows subjects' satisfaction rating with respect to issue 1
given the fact that they negotiated an agreement of one unit on this issue in both treatments ($s_1^R(1)$ and $s_1^N(1)$). The columns show whether
issue 1 was the respective subjects' reference issue (treatment 1) or
not (treatment 2). Median satisfaction is higher if issue 1 was the non-
reference issue, i.e. if the outcome is unexpectedly good. For an agreement of one unit on issue 1, a χ^2 test on homogeneity can just reject
the null hypothesis that in the underlying population represented by
the samples satisfaction is homogenous for subjects having issue 1 as
reference issue and subjects having issue 2 as reference issue (n = 64,
$\chi^2 = 9.478$, sim. p-value = 0.049).

Table 5.13. Contingency table on gain/loss utility given an outcome of one
unit on issue 1

	issue 1, agreement=1		
	$s_1^R(1)$	$s_1^N(1)$	Σ
very dissatisfied (1)	5	0	5
(2)	11	2	13
(3)	7	6	13
(4)	9	7	16
very satisfied (5)	7	10	17
Σ	39	25	64

Analogously, Table 5.14 gives the satisfaction ratings for an outcome
of 1 on issue 2. The trend is the same as for issue 1 and, again, there
is a significant difference (χ^2 test on homogeneity, n = 75, $\chi^2 = 9.887$,
sim. p-value = 0.036).

Table 5.14. Contingency table on gain/loss utility given an outcome of one unit on issue 2

	issue 2, agreement=1		
	$s_2^R(1)$	$s_2^N(1)$	$\sum$
very dissatisfied (1)	3	0	3
(2)	9	4	13
(3)	13	5	18
(4)	9	14	23
very satisfied (5)	7	11	18
$\sum$	41	34	75

RESULT 5.9: *Subjects satisfaction rating with an agreement of one unit on issue 1 or 2 significantly favors the assumption of gain/loss utility.*

Table 5.15 displays the subjects' report on how satisfied they are with an outcome of zero units on issue 1. In treatment 1, where issue 1 is the reference issue, agent A offers exactly one unit of issue 1 in each single offer. Furthermore, the arbitrator selects an agreement with exactly one unit of issue 1 if the two negotiators do not reach an agreement. Thus, the only way how a subject in treatment 1 can end up with zero units of issue 1 is if she makes an offer claiming these zero units and the agent accepts the offer. This happened only twice, as can be seen from Table 5.15. It is somewhat surprising that these two subjects were 'very satisfied' with not getting anything on issue 1. However, the surprise is lessened by the fact that they did not ask for more during the negotiation.

Table 5.15 suggests that subjects who get zero units on issue 1 are more satisfied if this was their reference issue than when it was there non-reference issue. This is, however, highly susceptible due to the small sample size, especially for the reference issue. Consequently, for an agreement of zero on issue 1, a χ^2 test on homogeneity cannot reject the null hypothesis that in the underlying population represented by the sample satisfaction is homogenous in the groups that have issue 1 as their reference issue or their non-reference issue (χ^2, n = 18, χ^2 = 3.536, sim. p-value = 0.405).

Finally, the analysis of subjects who negotiated an agreement of zero units on issue 2 is remaining. However, no subject in treatment 2 ended up with zero units on issue 2. Thus, there are no satisfaction ratings

Table 5.15. Contingency table on gain/loss utility given an outcome of zero units on issue 1

	issue 1, agreement$=0$		
	$s_1^R(0)$	$s_1^N(0)$	Σ
very dissatisfied (1)	0	4	4
(2)	0	3	3
(3)	0	3	3
(4)	0	1	1
very satisfied (5)	2	5	7
Σ	2	16	18

to compare with the ratings' by the seven subjects that had issue 2 as there non-reference issue and negotiated an agreement of zero units.

RESULT 5.10: *For an agreement of zero units on either issue 1 or issue 2, no significant influence of the reference issue on subjects' satisfaction can be shown.*

Summary

Overall, issues 1 and 2 both favor the notion of consumption utility (for issue 2 this is significant, for issue 1 it is not). Gain/loss utility has to be analyzed for four different cases separately to guarantee independence of observations. Subjects' satisfaction given that they negotiated an agreement of one unit suggests that gain/loss utility is a second source of utility besides consumption utility. This is (just) significant for issue 1 and 2 alike. For an agreement of zero units, there is no significant difference for issue 1 due to the small sample size and no comparison at all for issue 2 due to the very small sample size, i.e. zero observations from treatment 2.

Consumption utility is assumed by the rational choice model as well as by the attachment effect model and, thus, the data is in line with predictions by both models. The analysis of gain/loss utility, on the other hand, favors the attachment effect model and cannot be explained by the rational choice model.

5.3.4 Preference Uncertainty

According to the attachment effect model, subjects choosing their reference product should tend to have lower preference uncertainty than subjects choosing their non-reference product (cf. Sec. 5.1.5). In the following, two different proxies for preference uncertainty in choosing one of agent C's offers are used: response time and a self report on the complexity of choice.

The rational choice model does not include the notion of preference uncertainty. Thus, the implicit assumptions are (1) that response time might vary due to cognitive ability, chance, etc. but not depending on the reference or non-reference product and (2) that complexity of a choice is independent of the outcome of the choice, as choosing simply means retrieving preferences from memory and acting accordingly.

Association of the Proxies

Subjects' response times and the complexity of choice they report are two different measures for one underlying concept: preference uncertainty. For both measures, high values favor the assumption that the subject has high preference uncertainty. Thus response times and reported complexity should be positively correlated.

Given a single set of information and a specific task to perform, e.g. choice of an offer by agent C, several factors might influence a subject's response time—most prominently these are the cognitive and physical ability, the attention the subject pays to the computer screen, her familiarity with the handling of the computer, chance, and preference uncertainty. The latter is the only factor that differs systematically between choosing the reference and the non-reference product.

Subjects enter the choice of an offer by agent C and the complexity rating on the same screen. Thus, response time measures not only the time for choosing but as well the time for judging complexity. However, it is assumed that the time for rating the complexity is not systematically related to choosing the reference or the non-reference product. Furthermore, in the following discussion it is assumed that (most) subjects rate the complexity after choosing, i.e. they process the information and choices on the screen from top to bottom.

Response time and reported complexity are significantly positively correlated (one-sided correlation test, Spearman's rank correlation, $r_S = 0.221$, n = 82, S = 71532, p-value = 0.023). Thus, the two proxies measure at least partially the same.

The correlation can as well be measured for the choice of the dominant alternative offered by agent B as subjects were asked for a complexity rating here as well. For this choice too, response time and reported complexity are significantly positively correlated (one-sided correlation test, Spearman's rank correlation, $r_S = 0.229$, n = 82,

S = 70866, p-value = 0.019). The similarity of correlations for different choices strengthens the supposition that both proxies are related to the same underlying concept.[21]

The association r_S of the two proxies is not very strong. Possible explanations like the different scale levels were outlined in Section 5.1.5. Nevertheless, both proxies measure roughly the same and the results corroborate one another. Response time is, however, the more reliable measure as it applies to the choice itself and not the ex-post judgment of complexity that might be confounded as suggested by self-perception (Bern, 1967, 1972) and cognitive dissonance theory (Festinger, 1957; Festinger and Aronsons, 1960).

Response Time

Response time for choosing one offer by agent C ranged from 9 to 70 seconds with a median of 23 seconds (mean: 25.5 seconds). Figure 5.11 displays response times of subjects with respect to whether they choose the reference or the non-reference product.

The attachment effect model suggests that subjects choosing the reference product (R) and not the non-reference product (N) have lower preference uncertainty and are thus faster ($t^R < t^N$) and the rational choice model predicts no systematic difference ($t^R = t^N$, cf. Sec. 5.1.5). The tendency is in the direction suggested by the attachment effect model and the difference is significant; one-sided Wilcoxon rank sum test on the null hypothesis that in the underlying populations represented by the samples response time is not lower for subjects choosing the reference product than for subjects choosing the non-reference product (continuity correction and normal approximation, $n_1 = 55$, $n_2 = 27$, W = 501, p-value = 0.009).

RESULT 5.11: *Subjects choosing the reference product are significantly faster in doing so than subjects choosing the non-reference product.*

Reported Complexity

In the attachment effect model, a subject choosing the reference product on average has lower preference uncertainty and thus reports lower

---

[21] Although the correlation of the proxies in deciding on the offer by agent B is significant, it should not be over-interpreted. 75 out of 82 subjects judged this choice as 'very easy'. Given the dominance relation of the offers under consideration, this is plausible. However, there is hardly variation in the complexity and, thus, the correlation here is not too suggestive.

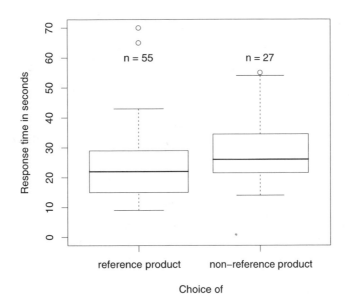

Fig. 5.11. Response times for the choice of an offer by agent C

complexity $(c^R < c^N)$ whereas in the rational choice model, there is not systematic difference depending on which product is chosen $(c^R = c^N,$ cf. Sec. 5.1.5).

Table 5.16 displays as how complex subjects judged the choice among offers by agent C depending on whether or not they choose the reference product. A χ^2 test on independence can just not reject the null hypothesis that complexity and choice of the reference or non-reference product are independent of one another in the underlying population represented by the sample (n = 82, χ^2 = 9.001, sim. p-value = 0.057). Thus, the tendency of reported complexity corroborates the finding of the above analysis of response times but the hypothesis test cannot confirm the significance of this difference.[22]

---

[22] Note, that a χ^2 test on contingency tables larger than 2 × 2 is inherently non-directional, whereas the Wilcoxon rank sum test applied to response times is directional. Thus, in general a higher p-value for the complexity is not surprising (about twice as high would be expected). A p-value about six times higher, on the other hand, is.

Table 5.16. Contingency table on the reported complexity of choice among offers by agent C

	Choice of		
	reference product	non-reference product	$\sum$
very easy (1)	20	8	28
(2)	21	4	25
(3)	8	8	16
(4)	4	3	7
very complex (5)	2	4	6
$\sum$	55	27	82

RESULT 5.12: *Subjects choosing the reference product tend to report a lower complexity of this choice than subjects who choose the non-reference product. The difference is, however, not significant.*

Summary

Two proxies were used to assess preference uncertainty in the choice among offers by agent C. Response time of subjects is seen as the more reliable proxy—it favors the attachment effect model over the rational choice model. The subjects' self-report on complexity exhibits a non-significant trend in the same direction. Overall, the attachment effect model organizes the data on preference uncertainty better than the rational choice model does.

5.3.5 Summary of Non-Parametric Results

To conclude the non-parametric analysis, the main results derived so far are summarized. The overview on the data produced the following findings (Sec. 5.3.1).

1. Subjects choose products 1 and 2 independently of one another and in both categories one of the products is chosen more frequently than the other.
2. In the alternating offer negotiation, subjects use different negotiation tactics and patterns thereof with approximately the same frequency as agent A does.
3. Subjects and the agent alike tend to make concessions and gradually approach the counterparty's ideal outcome with increasing offer numbers.

Based on these results that favor (1) the assumption of positive marginal utility for both products independently and (2) the assumption that the software agent in the alternating offer negotiation does not behave too different than a human negotiator would, three measures for an attachment effect are applied. The first—binary choice of the preferred product—gives the following results (Sec. 5.3.2).

4. Offers in the alternating offer negotiation have a significant systematic influence on the subjects' preferences. As predicted by the attachment effect model, subjects tend to choose the product that was offered more frequently.
5. The length of a negotiation with agent A has an impact on choosing the reference product or not. Subjects with a high number of offers tend to choose the reference product.
6. The proportion of male subjects choosing the reference product is higher than the proportion of female subjects choosing the reference product. The difference is, however, not significant.
7. The preference for the reference product cannot be explained by a significant influence of gender, age, course of studies, experience with experiments, experience with negotiation experiments, choice of two specific products at the beginning of the experiment, or different sessions.

Binary choice is seen as the most direct measure of subjects' ex-post preferences: it provides strong evidence for an attachment effect. This robust result is corroborated by a second set of statistics on the sources of utility (Sec. 5.3.3).

8. issues 1 and 2 favor the assumption that subjects satisfaction reflects consumption utility—for issue 2 this is significant, for issue 1 it is not.
9. Subjects satisfaction rating with an agreement of one unit on issue 1 or 2 significantly favors the assumption of gain/loss utility.
10. For an agreement of zero units on either issue 1 or issue 2, no significant influence of the reference issue on subjects' satisfaction can be shown.

The data suggests that there are (at least) two sources of utility: consumption utility and gain/loss utility. Again, these findings favor the attachment effect model over the rational choice model. Yet another way to assess an attachment effect is via the subjects' preference uncertainty. The results concerning this measure are as follows (Sec. 5.3.4).

11. Subjects choosing the reference product are significantly faster in doing so than subjects choosing the non-reference product.
12. Subjects choosing the reference product tend to report a lower complexity of this choice than subjects who choose the non-reference product. The difference is, however, not significant.

The two different proxies used for preference uncertainty—response time and reported, subjective complexity of choice—hint in the same direction (although just on of them significant): negotiators choosing their respective reference product exhibit lower preference uncertainty than negotiators choosing the non-reference product. Again, these results suggest that the attachment effect model organizes the data significantly better than the more restrictive rational choice model. So far, the analysis only used qualitative, non-parametric predictions derived from the attachment effect model. Next, it turns to an estimation of the update functions to quantify the effect of single offers on reference points.

5.4 Parametric Analysis

To provide external validity for the theoretical model of the attachment effect and determine empirically the size of the effect, data from the laboratory experiment is employed to estimate parameters. For this, the version with linear update functions that was introduced in Section 3.3.3 is used for two reasons: Firstly, rational choice is an instance of this model with a given parameterization. Thus, the estimation of parameters automatically becomes a test whether the attachment effect does a significantly better job at organizing the data than rational choice does. Secondly, the model uses a minimum of parameters to allow (1) for all functions to vary depending on the difference of offers and reference points and (2) for the effect being differently pronounced depending on the party that makes the offer and on the direction of adjustment. Keeping the number of parameters low is essential for estimating them reliably with the data at hand. That such a relatively simple, linear model is sufficient will be seen from the results.

The model is given as follows for all $k \in \{1, 2\}$: The initial reference point is

$$r_k^0 = \underline{x}_k + \beta_0(\overline{x}_k - \underline{x}_k)$$

with $\beta_0 \in [0, 1]$, $\underline{x}_k$ as the lower limit of the agreement space on issue k, and $\overline{x}_k$ as the upper limit of the agreement space on issue k. As the agreement space is $\{0,1,2\}$ for both issues, this can be written as $r_k^0 = 2\,\beta_0$. Subsequently, the reference point is defined recursively as

$$
r_k^t = \begin{cases}
r_k^{t-1} + \beta_1(x_k^t - r_k^{t-1}) & \text{if } t = 1,3,\cdots \text{ and } x_k^t \geq r_k^{t-1}, \\
r_k^{t-1} - \beta_2(r_k^{t-1} - x_k^t) & \text{if } t = 1,3,\cdots \text{ and } x_k^t < r_k^{t-1}, \\
r_k^{t-1} + \beta_3(x_k^t - r_k^{t-1}) & \text{if } t = 2,4,\cdots \text{ and } x_k^t \geq r_k^{t-1}, \\
r_k^{t-1} - \beta_4(r_k^{t-1} - x_k^t) & \text{otherwise}
\end{cases}
$$

with $\beta_1, \beta_2, \beta_3, \beta_4 \in [0,\ 1]$.

Each subject contributes a single observation to the data set. Thus, it is not possible to estimate the strength of the attachment effect for individual subjects. Instead, the estimation pools the data and finds an 'average' effect, i.e. the parameters are not identified for signal subjects. Only issues 1 and 2 are considered here, as the choice of offers by agent C does not allow any inference on the reference point with respect to issue 3. Furthermore, the data does not allow estimating the difference between issues 1 and 2 and is thus pooled in this dimension as well. As the goods used as issue 1 and as issue 2 are relatively comparable, i.e. durable consumer goods with about the same retail price, no significant difference between issues would be expected anyways.

The agreement space imposed by the experiment design is discrete. However, the reference point here is allowed to shift continuously within the bounds of the agreement space for two reasons: Firstly, the reference point is assessed for a set of subjects and not for each individual separately and, secondly, even for an individual negotiator the reference point might not be restricted to discrete steps. Thus, the initial reference point is assumed to be an element of the convex hull of the agreement space (denoted as X'), the update functions exhibit all characteristics proposed as reasonable in Section 3.3.2, and the different parameters allow to compare the strength of the attachment effect depending on the nature of the offer. Rational choice is included as special case with $\beta_1 = \beta_2 = \beta_3 = \beta_4 = 0$.

5.4.1 Foundations of the Estimation

The maximum likelihood method is used for estimation of β_0, β_1, β_2, β_3, and β_4 simultaneously. Maximum likelihood estimators determine the set of parameters under which the sampled data has the highest likelihood of being observed. This method was already introduced by Fisher (1921, 1925) and gained greater importance since computers are available for data analysis. See e.g. Cramer (1986) for an introduction of maximum likelihood methods in econometrics.

Likelihood in Individual Binary Choice

Given the offers observed in the experiment, a subject's reference point at the end of negotiating with agent A can be calculated for any set of model parameters. The duration of subject i's negotiation is denoted as $T_i \in \{1, 2, \cdots, 12\}$ and her hypothetical reference point at the end of negotiation is calculated on basis of the model and denoted as $r_i^{T_i} \in X'$. The subject's true reference point is, however, not observed and can thus not be directly compared to the hypothetical one. Hence, another measure has to be found to assess the likelihood of a parameter combination. This will be the binary choice of an offer by agent C. For this, the three-dimensional[23] reference point has to be transformed to a one-dimensional likelihood of choosing either offer.

It is assumed that a subject's true reference point influences her choice among offers by agent C. A subject with a high reference point on issue 1 and a low reference point on issue 2, for example, perceives a (stronger) loss when choosing issue 2 than when choosing issue 1. Thus, she is more likely to choose agent C's first offer. On the contrary, a subject with a high reference point on issue 2 and a low reference point on issue 1 will tend to choose the agent's second offer. No clear prediction of a subject's propensity to choose either offer can be made when the reference point is about equal on both issues.

The relation of the reference point and choice of one of C's offers can be formalized as

$$p(c_i = 1 | r_i^{T_i} = \langle r_{i,1}, r_{i,2}, \cdot \rangle) = \frac{r_{i,1}}{r_{i,1} + r_{i,2}}$$

where the binary indicator variable $c_i \in \{0, 1\}$ stands for the subject choosing offer 1 ($c_i = 1$) or offer 2 ($c_i = 0$) and $p(\cdot)$ for the probability of choosing an offer conditional on a reference point. The location of the reference point on product 3 is assumed to be irrelevant for the choice of either product 1 or product 2. With the above calculation, $p(c_i = 1 | r_i^{T_i} = \langle 0, 0, \cdot \rangle)$ is not defined; it is assumed to be 0.5, as not directed prediction for the choice can be made. For all other $r_i^{T_i} \in X'$, the probability is defined by the above formula.

---

[23] Although issue 3 is not considered in the model, the reference point per se is three-dimensional.

This relatively simple function captures the intuition outlined before: Given a fix reference point on product 2, an increase of the reference point on product 1 increases the probability of choosing product 1. An equal reference point on both issues leads to the probability being 50% for choosing either offer. Furthermore, the probability $p(\cdot)$ is bound to the range $[0,\ 1]$ with the agreement space specified in the experiment design.

The likelihood l_i of observing a specific choice c_i by subject i is then given as

$$l_i = \begin{cases} p(c_i = 1 \mid r_i^{T_i}) & \text{if } c_i = 1, \\ 1 - p(c_i = 1 \mid r_i^{T_i}) & \text{otherwise} \end{cases}$$

which can be written as $l_i = (2c_i - 1)\, p(c_i = 1 \mid r_i^{T_i}) + (1 - c_i)$ without case differentiation. With the observed offers in the negotiation, the recursive definition of the reference point $r_i^{T_i}$, and an observed choice c_i, the likelihood is a function of the model parameters.

Besides the negotiation with agent A, the offer by agent B and the offers by C likely have an influence on a subject's reference point. However, as (1) these offers do not differ between subjects, (2) the offer by B is the same on issues 1 and 2, and (3) the offers by C interchange the values on issues 1 and 2, it is assumed that these offers do not introduce a systematic differentiation between subjects and do not change the choice of either offer by agent C in a specific direction. Hence these offers are omitted for simplicity.

Global Likelihood

Each of 82 subjects contributes and independent observation. Thus, the overall likelihood L of observing exactly the 82 choices made by the subject is the product of the individual likelihoods:

$$L(\beta_0, \beta_1, \beta_2, \beta_3, \beta_4) = \prod_{i=1}^{82} l_i = \prod_{i=1}^{82} (2c_i - 1)\, p(c_i = 1 \mid r_i^{T_i}) + (1 - c_i)$$

where for each subject $r_i^{T_i}$ is a function of the data (the offers recorded in the alternating offer negotiation) and the model parameters (β_0, β_1, β_2, β_3, β_4); c_i is the agent's observed choice. The maximum likelihood estimation of the update parameters is obtained by maximizing the likelihood function $L(\beta_0, \beta_1, \beta_2, \beta_3, \beta_4)$ over the set of all possible parameter combinations.

Numerical Estimation

In the absence of an explicit analytical solution to this optimization, the likelihood function is maximized numerically. For this, an undirected grid search over the entire five-dimensional parameter space is performed.[24]

Initially, the likelihood function was calculated with a grid of step size 0.05 on all dimensions. In a second step, the process was repeated with a finer grid around the maximum found so far. The second grid was applied to a subspace of ±0.05 of the maximum so far with a step size of 0.01 on each dimension. In a third step, an even finer grid with a step size of 0.001 was used for the subspace of ±0.01 around the maximum found so far. Finally and for each parameter individually, the likelihood was calculated for a step size of 0.001 over the entire parameter range (i.e. $[0, 1]$ for each parameter) conditional on the maximum likelihood estimates of the other four parameters.

Confidence Intervals

Confidence intervals (CI) for parameter estimates can be calculated by means of the likelihood ratio method as described by, e.g., Agresti (1990, Ch. 1). This method is commonly applied to test significance of maximum likelihood estimates and to construct confidence intervals. To test whether the maximum likelihood estimate $\hat{\beta}_j$ and an alternative parameter value β'_j differ significantly, a likelihood ratio test assumes that under the null hypothesis of no systematic difference twice the (negative) logarithm of the ratio of likelihoods at these two parameter values follows an asymptotic χ^2 distribution with one degree of freedom. Formally this means that $LR = -2\log\left(\frac{L(\beta'_j,\hat{\beta}_{-j})}{m}\right) \sim \chi^2_1$ where $m = L(\hat{\beta}_j, \hat{\beta}_{-j})$ is the overall maximum likelihood and $L(\beta'_j, \hat{\beta}_{-j})$ is the likelihood for β'_j conditional on the maximum likelihood estimates of all other parameters. The notation $\hat{\beta}_{-j}$ stands for the maximum likelihood estimates of all parameters except β_j.[25]

Confidence intervals are constructed by inverting the test. For β_j, the 95% confidence interval is the set of all β'_j for which the test cannot reject the null hypothesis $\beta_j = \beta'_j$ at a 5% level. Confidence intervals for

---

[24] See e.g. Cramer (1986, Sec. 5) for more sophisticated directed search techniques requiring less computations. However, given the computing power available and the problem size, an undirected 'brute force' search appears appropriate.

[25] By construction $L(\hat{\beta}_j, \hat{\beta}_{-j}) \geq L(\beta'_j, \hat{\beta}_{-j}) \geq 0$ holds. Thus, the quotient is bound to the interval $[0, 1]$, its logarithm is non-positive, and the test statistic LR is non-negative.

single parameters are conditioned on the maximum likelihood estimates of the other parameters. With the maximum likelihood m and a value of 3.84 for the 5% right tailed probability of a χ^2 distribution with one degree of freedom, the confidence interval of a parameter β_j is defined as

$$CI_{0.95}(\beta_j, \hat{\beta}_{-j}) = \{\beta'_j \mid L(\beta'_j, \hat{\beta}_{-j}) \geq m \, e^{-1.92}\}$$

These intervals can directly be derived from the calculations made for point estimation.

5.4.2 Maximum Likelihood Estimation

The numerical maximization lead to a unique maximum: Point estimates and 95% confidence intervals of the parameters are given in Table 5.17. Furthermore, for each parameter individually, a likelihood ratio test on the null hypothesis that the parameter is equal to zero is performed. The test is conditioned on the other parameters' estimates and the test statistic LR is assumed to follow a χ^2_1 distribution. Values of the test statistic and p-values are given in the table.

Table 5.17. Maximum likelihood estimate of parameters in the linear attachment effect model

Parameter	ML estimate $\hat{\beta}_j$	$CI_{0.95}(\beta_j, \hat{\beta}_{-j})$	Test on $H_0 : \beta_j = 0$ LR	p-value
β_0	0.748	[0.286 , 1]	inf.	< 0.001
β_1	0.116	[0.003 , 0.497]	4.399	0.036
β_2	0.286	[0.16 , 0.398]	14.45	< 0.001
β_3	0	[0 , 0.058]	0	> 0.5
β_4	0.523	[0.22 , 0.768]	10.854	0.001

Confidence intervals and likelihood ratio tests suggest that four of the five parameters are significantly different from zero, i.e. significantly different from the assumption underlying the rational choice model. For each single parameter, this result is conditioned on the estimates of the other parameters. However, its robustness is corroborated by a combined test of all parameters simultaneously. A likelihood ratio test can reject the restrictive null hypothesis $\beta_1 = \beta_2 = \beta_3 = \beta_4 = 0$ in favor of the more general alternative hypothesis $\beta_1, \beta_2, \beta_3, \beta_4 \in$

$[0, 1]$ (test statistic $LR = -2\log\left(\frac{L(\cdot,0,0,0,0)}{m}\right) = 27.367$, χ_4^2 distribution, p-value < 0.001).[26] Thus, the estimation allows rejecting the applicability of the rational choice model for the data set and suggests that the attachment effect model is better suited to explain the observed choices by negotiators.

RESULT 5.13: *A maximum likelihood estimation allows quantifying the effect of single offers on reference points.*

RESULT 5.14: *The individual parameter estimates as well as the overall model fit show that the attachment effect model organizes the data significantly better than a rational choice model.*

Interpretation

The exact values of the estimate should be interpreted cautiously as the confidence intervals are rather wide. $\hat{\beta}_1$, for example, is less than half of $\hat{\beta}_2$. Both parameters relate to the adaptation of the reference point when a negotiator receives an offer from her counterparty. Thus, the estimation suggests that an upward adjustment ($\hat{\beta}_1$) of the reference point is by far slower than a downward adjustment ($\hat{\beta}_2$). However, as the confidence interval for β_2 is a subset of the interval for β_1, the difference between the two point estimates might be by chance.

An initial reference point of 74.8% of the possible range in the agreement space, i.e. 1.496 units on a scale from 0 to 2 units, appears rather high. However, the value is in line with commonly reported overconfidence prior to a negotiation that leads to excessively optimistic judgments about the likelihood of getting a good outcome (cf. Sec. 3.1.3; Kramer, Newton, and Pommerenke, 1993; Lim, 1997).

---

[26] The parameter β_0 (the initial reference point) is not decisive for the likelihood under the null hypothesis. Given the rational choice model, the initial reference point will be the final reference point. As it has the same value on all issues, the likelihood of choosing either offer by agent C will be 50% for each single subject ($\forall\, i \in \{1, 2, \cdots, 82\} : p(c_i = 1 | r_i^{T_i} = r_i^0) = 0.5$). Thus, $L(\cdot, 0, 0, 0, 0)$ is invariant to β_0.

A more restrictive null hypothesis would additionally require $\beta_0 = 0$. The interpretation is that the reference point is not only constant over time but it is equal to the worst possible agreement in the agreement space and, thus, all agreements are evaluated as gains. While this does not change the test statistic, it would require a comparison with a χ_5^2 distribution. Such an increase in the degrees of freedom generally makes it more difficult to reject the null hypothesis. Here however, the result is robust to such a change and the p-value remains well under 0.001.

RESULT 5.15: *At the beginning of a negotiation, subjects are rather optimistic in the agreement they might achieve.*

From this overly optimistic initial reference point on, downward adjustments are stronger than upwards adjustments ($\hat{\beta}_2 > \hat{\beta}_1$ and $\hat{\beta}_4 > \hat{\beta}_3$). Overall, downward adjustments based on a negotiator's own offers are strongest, followed by downward adjustments based on the counterparty's offers. This can be interpreted as negotiators quickly adjusting their over optimism during the negotiation. If the counterparty does not offer much on one issue, than the expectation in the outcome is reduced. If the negotiator herself does not even demand to get much on that issue, it appears that this diminishes her expectation and hence her reference point even stronger.

For upward adjustments, the relative strength of own offers and offers by the counterparty is the other way round: while offers by the counterparty significantly increase the reference point ($\hat{\beta}_1 = 0.116$), a negotiator's own offers do not increase her reference point. It seems that the fact that a negotiator demands an agreement does not increase her expectation in really obtaining it while the fact that it is offered by the other party increases her reference point even if the negotiation continues. The estimates appear plausible. However, the ranking might partially be due to chance as the confidence intervals suggest.

RESULT 5.16: *A negotiator's own offer can lead to a strong decrease of her reference point. It does, however, not increase her reference point significantly.*

RESULT 5.17: *A negotiator's reference point can be influenced by her counterparty's offer. Increases of the reference point thereby seem to be weaker than decreases.*

Hypothetical Reference Points

Figure 5.12 shows the hypothetical reference points of the 82 subjects that can be calculated given the maximum likelihood estimation. Circles represent subjects in treatment 1: solid circles are thereby used for subjects that chose the reference product, i.e. product 1 ($c_i = 1$), and non-solid circles for subjects that chose the non-reference product, i.e. product 2 ($c_i = 0$). Analogously, solid squares represent subjects in treatment 2 that chose the reference product, i.e. product 2 ($c_i = 0$), and non-solid squares represent subjects that chose the non-reference product, i.e. product 1 ($c_i = 1$). Thus, solid entries represent subjects that chose as predicted by the attachment effect which is the majority of entries in the figure (55 out of 82).

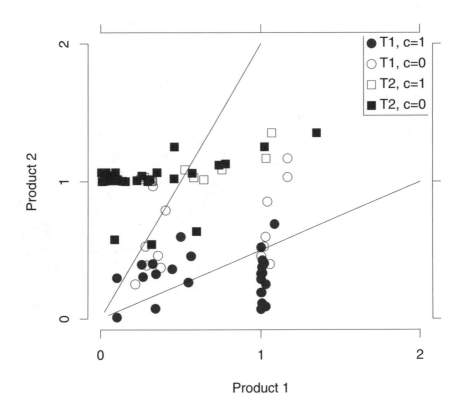

Fig. 5.12. Reference points given the maximum likelihood estimates

The upper straight line in Figure 5.12 indicates all reference points r for which $p(c = 1|r) = \frac{1}{3}$ and the lower line for all reference points with $p(c = 1|r) = \frac{2}{3}$. The figure suggests two patterns for reference points: Firstly, reference points from subjects in treatment 1 (the circles) tend to be in the lower right part of the figure and reference points from subjects in treatment 2 (the squares) tend to be in the upper left part. Thus, the treatments have achieved their objective of directing reference points to different parts of the agreement space by employing different patterns in the offer sequences.

Secondly, most subjects with reference points below the lower straight line, i.e. subjects with $p(c_i = 1|r_i) \geq \frac{2}{3}$, chose product 1. Analogously, most subjects with reference points above the upper straight line, i.e. subjects with $p(c_i = 1|r_i) \leq \frac{1}{3}$, chose product 2. Thus, the majority of subjects for which the attachment effect model makes a rather precise prediction on what they are likely to choose really behaved as predicted. On the contrary, there is no clear pattern for subjects for whom the attachment effect model predicts roughly the same probability of choosing either product (entries between the two lines).

5.4.3 Reliability of the Estimation

The parameter estimates given in Table 5.17 appear reasonable for three reasons studied so far:

1. The confidence intervals are strict subsets of the possible parameter ranges and the overall estimate is significantly better than the rational choice model. Thus, differences of likelihoods in the parameter space are not solely by chance but the estimate is significantly better than infinitely many other parameter combinations.
2. The estimates can easily be interpreted and the relation to a negotiator's expectations appears plausible.
3. The subjects' reference points calculated from the estimated model are diverse and show patterns that correspond to the subjects' choices.

To further strengthen the reliability of the estimation, three additional aspects are considered in this section: Firstly, properties of the likelihood function are studied and, secondly, the binary choice among offers by agent C is compared to the model's prediction. Both aspects assess the estimation's internal validity. Thirdly, the model is applied to data from the internet experiment to test its external validity. These three additional measures support the reliability of the estimation.

Conditional Likelihood Functions

Figure 5.13 plots the likelihood of observing the data as function of the individual parameters.[27] In each of the plots one parameter varies while the other four are fixed at their estimates. Figure 5.13a, for example, shows the dependence of the likelihood function on β_1 with a maximum at $\hat{\beta}_1 = 0.116$. The horizontal straight line represents the likelihood level $e^{-1.92} = 14.661\%$ obtained from the likelihood ratio test for constructing confidence intervals, i.e. each parameter value that gives a likelihood at least this high falls in the confidence interval.

---

[27] In fact the ordinate is not the likelihood itself but the likelihood as percentage of its maximum.

The purpose of Figure 5.13 is to assess characteristics of the likelihood function. In the absence of an analytical solution to the maximization, the plotted functions indicate that all conditional likelihood functions are single peaked and steady. This strongly supports the supposition that the numerically determined estimates reported in Table 5.17 indeed constitute the global maximum likelihood estimates.

To study interaction of pairs of parameters, Figure 5.14 (displayed on pages 218 and 219) plots the likelihood surfaces for each pair of parameters, i.e. the likelihood as function of two of its parameters conditional on the estimates of the other three parameters. Like Figure 5.13, Figure 5.14 shows that the conditional likelihood functions are single peaked and steady and thus further corroborates the assumption that the parameter values given in Table 5.17 define the global maximum likelihood.[28]

RESULT 5.18: *The likelihood function conditioned on any single parameter or any pair of parameters is steady and single-peaked. Thus, the numerical estimation likely found the true maximum likelihood.*

Binary Choice

Given the reference point for each subject, the probability of choosing product 1 ($p(c_i = 1 \mid r_i^{T_i})$) that was used in the estimation can be calculated for each subject. The estimation determined the set of parameters for which the likelihood of the observed choices is highest. Thus, the estimators should be determined so that the model predicts a high likelihood of choosing product 1 for subjects that indeed chose product 1 and a low probability for subjects that chose product 2. In other words, finding the maximum likelihood of the data is the same as differentiating the two groups of subjects 'as good as possible' in terms of probabilities predicted for them. Thus, the distributions of probabilities should differ significantly for both groups if the estimation is reliable.

The distributions of estimated probabilities for the two groups of subjects that either chose product 1 or product 2 are displayed in Figure 5.15. The values are calculated from the offers exchanged in the alternating offer negotiation and the estimated model. The figure clearly

---

[28] Figure 5.14h plots multiple local maxima for the likelihood function with respect to β_0 and β_2. This is, however, due to the grid width used for plotting. With a resolution of 0.001 on both parameters, the function has only a single local maximum which thus is the global maximum.

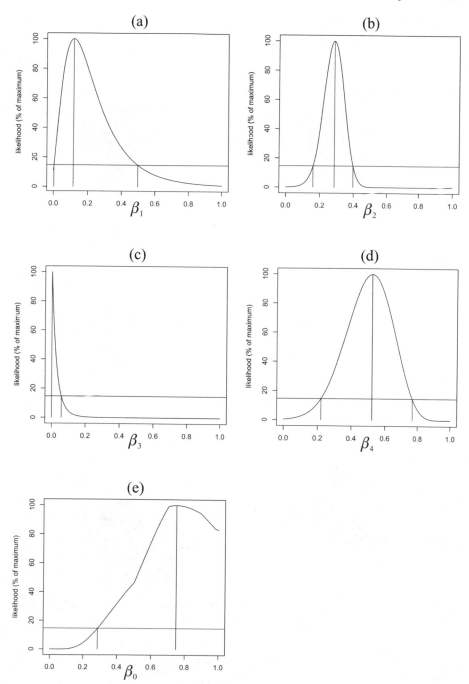

Fig. 5.13. Conditional likelihood functions for single parameters

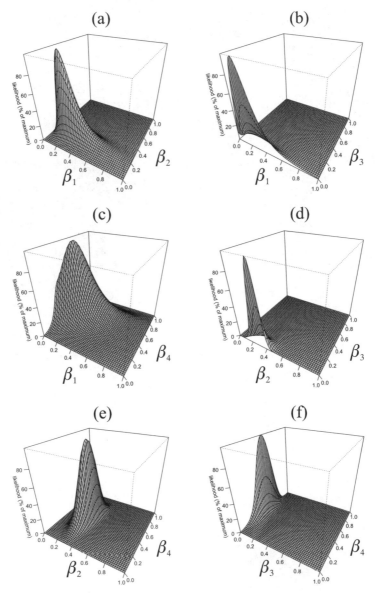

Fig. 5.14. Conditional likelihood functions for pairs of parameters (Part 1 of 2)

(g)   (h)

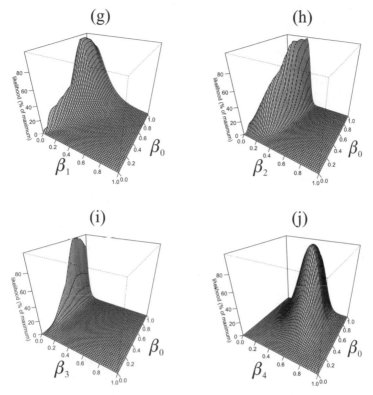

(i)   (j)

Fig. 5.14. continued ... (Part 2 of 2)

suggests that the estimated probability for subjects that chose product 1 (38 subjects) tends to be higher than for the subjects that chose product 2 (44 subjects). The smallest probability of choosing product 1 calculated for any subject that really chose product 1 is, for example, 19.8%. On the contrary, the estimated probability is lower than this 19.8% for more than one third of subjects that chose product 2.

The significance of the expected difference can be tested via a one-sided Wilcoxon rank sum test on the null hypothesis that in the underlying populations represented by the samples the probabilities for choosing product 1 estimated on basis of the offers exchanged is not higher for subjects that choose product 1 than for subjects that chose product 2. The null hypothesis can be rejected (continuity correction and normal approximation, $n_1 = 38$, $n_2 = 44$, $W = 1324$, p-value < 0.001).

Fig. 5.15. Distributions of estimated probabilities of choosing product 1 given the reference points

RESULT 5.19: *The estimated model predicts significantly higher probabilities for choosing product 1 for subjects that really chose product 1 than for subjects that chose product 2.*

External Validity: Data from the Internet Experiment

The ultimate test whether the estimates for the update functions are reliable estimates for the attachment effect is an out of sample test, i.e. comparing the estimated model with data that was not used in the estimation. To this end, the estimated model is applied to data from the internet experiment to see whether it can predict the subjects' *WTA* values (cf. Ch. 4).

Like in Section 4.1.5, the ordinal labels *no* and *yes* for issues *A* (elevator) and *B* (balcony) are coded as zero and unity. With the linear attachment effect model and the maximum likelihood estimates of its parameters, the expected reference point can be calculated for each

subject in the internet experiment. The reference points are given in
Figure 5.16. This plot closely relates to Figure 4.4 (p. 129) that was used
to derive a hypothesis for the internet experiment. The difference is that
in Section 4.1.5, the simplifying assumption $\forall\, a \in \mathbb{R}_+ : f_k^+(a) = \beta_1\, a$,
$f_k^-(a) = 0$, $g_k^+(a) = 0$, $g_k^-(a) = 0$, and $r_k^0 = 0 \ \forall\ k \in \{A, B\}$ with
$\beta_1 \in (0,\ 1]$ were made, whereas here the functions are based on data
from the lab experiment.

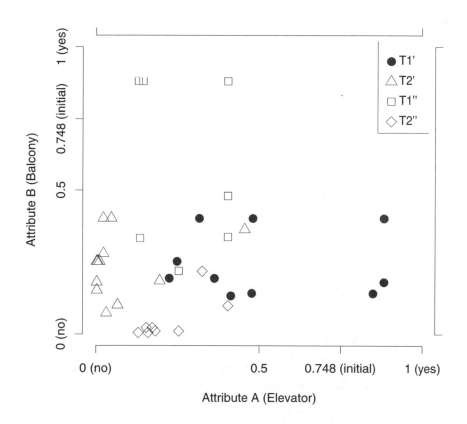

Fig. 5.16. Expected reference points in the internet experiment given the
maximum likelihood estimates from the lab experiment

As it is no longer assumed that most update functions are equivalent
to zero, reference points reflect more aspects of the alternating offer
negotiation like the subjects' own offers, for example. Consequently,

there is greater heterogeneity in the reference points. Graphically, this is reflected by less structure in Figure 5.16 than in Figure 4.4. The main patterns that were used to derive the hypothesis on a treatment effect do, however, persist: Subjects confronted with offer sequence *T1'* (solid circles) have a higher reference point on attribute *A* than subjects confronted with *T2'* (triangles), etc.

The simplifying ad-hoc assumptions in Section 4.1.5 only allowed to derive a hypothesis how the attachment effect might qualitatively be reflected in the data, i.e. $WTA_{Ref}^{T1} > WTA_{Ref}^{T2}$. Given the estimated model, the exact location of the reference point can now be expected to have a meaning. Thus, the general qualitative prediction of greater *WTA* values for some subjects than for others can be refined.

It is no longer necessary to differentiate treatments or offer sequences and to group subjects accordingly. Given a hypothesis on each subject's reference point, subjects with higher reference points on an issue should tend to report a higher *WTA* for this issue irrespective of any assignment to offer sequences. Hence, a positive correlation of reference points and *WTA*'s is expected for issues *A* and *B* individually.

The expected association of reference points and *WTA* values can be tested by a correlation test for each issue independently. As in Section 4.2, a rank-based test is applied to account for outliers and non-normality of *WTA* values. For each test, the null hypothesis is that there is no positive correlation between the subjects' reference point on an issue and her respective *WTA* on that issue. For both issues, the null hypothesis can be rejected (one-sided correlation test, Spearman's rank correlation, n = 42; attribute *A*: $r_S = 0.621$, S = 4671, p-value < 0.001; attribute *B*: $r_S = 0.648$, S = 4341, p-value < 0.001). Thus, there is a significant positive correlation between the reference points predicted by the estimated attachment effect model and the subjects' *WTA* values.

RESULT 5.20: *The parameter estimation has external validity: The estimated model allows the prediction of data gathered in the internet experiment.*

5.5 Summary of Results

To conclude the parametric analysis, the main results derived are summarized. The estimation of the linear attachment effect model led to the following results (Sec. 5.4.2):

- A maximum likelihood estimation allows quantifying the effect of single offers on reference points.
- The individual parameter estimates as well as the overall model fit show that the attachment effect model organizes the data significantly better than a rational choice model.
- At the beginning of a negotiation, subjects are rather optimistic in the agreement they might achieve.
- A negotiator's own offer can lead to a strong decrease of her reference point. It does, however, not increase her reference point significantly.
- A negotiator's reference point can be influenced by her counter-party's offer. Increases of the reference point thereby seem to be weaker than decreases.

The maximum likelihood point estimates for parameters of the attachment effect model with linear update functions (cf. Sec. 3.3 for the model and Table 5.17 for confidence intervals) are:

$$f^+(a) = 0.116\,a \qquad f^-(a) = 0.286\,a$$
$$g^+(a) = 0 \qquad g^-(a) = 0.523\,a$$
$$r^0 = \underline{x} + 0.748\,(\overline{x} - \underline{x})$$

With respect to the reliability of the estimation, the following results were obtained (Sec. 5.4.3):

- The likelihood function conditioned on any single parameter or any pair of parameters is steady and single-peaked. Thus, the numerical estimation likely found the true maximum likelihood.
- The estimated model predicts significantly higher probabilities for choosing product 1 for subjects that really chose product 1 than for subjects that chose product 2.
- The parameter estimation has external validity: The estimated model allows the prediction of data gathered in the internet experiment.

Thus, the estimation produced a reasonable and reliable quantification of the attachment effect and allows predicting the effect of single offers on reference points in the lab experiment as well as in the internet experiment. Overall, data from both experiments provide suggestive evidence for an attachment effect in multi-issue negotiations. The attachment effect model is thereby significantly better in assessing the negotiators' preferences than a rational choice model is. The attachment effect is a systematic bias that affects negotiators (at least in the two experiments)—this descriptive result can be used for prescriptive negotiation analysis, i.e. to advice negotiators in debiasing themselves and correctly anticipating their counterparty's behavior.

6

Conclusions and Future Work

There is an art and a science of negotiation. By "science" I loosely mean systematic analysis for problem solving [...] The "art" side of the ledger is equally slippery: it includes interpersonal skills, the ability to convince and be convinced, the ability to employ a basketfull of bargaining ploys, and the wisdom to know when and how to use them.

(Raiffa, 1982, pp. 7–8)

The study of offers, expectations, reference points, and attachment in negotiations classifies as 'science of negotiation'. It is concerned with understanding the cognition and behavior of negotiators and contributes to the descriptive foundations of negotiation analysis from which advice how to negotiate rationally can be deduced. Personal characteristics that belong to the 'art of negotiation', on the other hand, were not considered in the present work.

This chapter concludes the study at hand by firstly summarizing the work (Sec. 6.1). Thereby, the research questions, the main contributions, and the implications of the results are reviewed. Then, Section 6.2 critically discusses limitations of the present work, especially the question of external validity which is essential for any empirical work. Finally, Section 6.3 briefly outlines directions of future work to extend the present study.

6.1 Summary of Contribution and Review of Work

The focal question addressed is whether the offers exchanged in a multi-issue negotiation have a systematic effect on the negotiators' preferences. Such an effect—termed *attachment effect*—is derived from previous research in economics and psychology, modeled, and demonstrated empirically in the present study. In a traditional economic rational choice model such an effect is ruled out by the assumption that preferences are exogenously given and invariable. Behavioral economics (especially prospect theory) and cognitive psychology, however, suggest that preferences might be influenced by market processes, elicitation procedures, and the context of decision-making. The attachment effect draws on these behavioral perspectives on decision-making.

6.1.1 Contribution

The contribution of the present work is threefold:

1. It motivates and models an attachment effect in multi-issue negotiations and thereby contributes to the descriptive foundation of negotiation analysis and negotiation and market engineering.
2. The study empirically assesses the existence and the size of the attachment effect and thereby proves applicability of the attachment effect model.
3. The work contributes to the methodology of negotiation and market engineering by showing how the combination of two experiments can increase robustness and validity of results which is essential in an engineering context.

The present study has proceeded in several steps to accomplish the overall objective of studying the effect of offers on preferences and to achieve this contribution.

Chapter 1 motivated the present work, situated it within the research fields *negotiation analysis* and *negotiation and market engineering*, and outlined the research questions. A review of related work and an overview on the structure of the analysis completed the chapter.

Chapter 2 introduced and compared several theories on preferences and approaches to human decision-making starting with traditional microeconomic theory and then discussing behavioral economics, cognitive psychology, and neuro-sciences. The approaches differ with respect to their internal coherence, their congruence with reality, their abstraction, and their predominant research methodologies. Although there is no overall best approach, behavioral economics as mainly empirical field of research combining theoretical economics and psychological research was most fruitful for the subsequent study of the effect offers in a negotiation have on preferences.

Chapter 3 sketched the interdisciplinarity of research on negotiations and reviewed game theoretic and negotiation analytic approaches. As there are infinitely many Nash equilibria and (to date) no meaningful refinement of this equilibrium concept for bilateral multi-issue alternating offer negotiations under incomplete information, the overall work is presented in a negotiation analysis context. Process models adopted from information systems research were used to identify processes in which negotiators might be prone to biases. Based on this, the attachment effect in negotiations was defined and modeled. Chapter 4 presented an internet experiment which tests for the existence of the attachment effect in multi-issue negotiations between a human subject and a software agent controlled by the experimenter. The experiment provided suggestive evidence that the attachment effect model organizes data better than a rational choice model does. To increase internal and external validity of the results, refinements of the design were outlined at the end of the chapter.

Chapter 5 described a lab experiment that implemented the revised design. As the internet experiment, this second experiment favors the attachment effect model over a traditional rational choice model. The differences in the experimental designs proof that the result is robust and does not depend (too much) on design decisions. Furthermore, data from the lab experiment was used in a maximum likelihood estimation of parameters in the attachment effect model.

In the following, the theoretical background (Ch. 2 and 3), the empirical results (Ch. 4 and 5), and methodological issues concerning the experiments are reviewed in more detail.

Theory

The implications of reference points on preferences have been studied extensively over the last decades. The origin of reference points, on the other hand, is a grossly understudied topic; oftentimes the status quo is assumed to be the reference point. The present work analyses the formation and shift of reference points in the context of multi-issue negotiations. Understanding negotiator decision-making is thereby approached from a behavioral and an economic perspective and a model for the attachment effect in alternating offer negotiations is presented.

The attachment effect model assumes a causal relationship of

- offers made by the negotiators,
- their expectations in the outcome,
- issue-wise reference points,
- preferences, and
- choice.

The relation of offers and expectations is frequently used in, for example, game theoretic models on bargaining under incomplete information where offers signal preferences and influence the counterparty's beliefs. The effect of expectations on reference points has been noted repeatedly in prospect theory and was recently the core element of a model on reference-dependent preferences proposed by Köszegi and Rabin (2006). The effect of reference points on preferences over multiple issues has most notably been analyzed by Tversky and Kahneman (1991) and the interrelation of preferences and choice is standard in economics. Thus, each single step in the chain of causal relationships given above has already been addressed in previous research. The overall combination is, however, novel in this work. The traditional economic counter piece to the attachment effect model is a rational choice model that assumes preferences to be exogenously given and fix. Rational choice models might allow for the existence of a reference point and loss aversion but in any such model the reference point is assumed to be invariant to negotiation or other market processes.

Empiricism

The effect offers in negotiations have on preferences is derived theoretically in Chapters 2 and 3. Whether the attachment effect is present and reference-dependent preferences change systematically in multi-issue negotiations is then tested experimentally (Ch. 4 and 5). The two experiments are conducted to test applicability of the attachment effect model and a rational choice model. In the internet experiment, students negotiate on attributes of a hypothetical contract for renting an apartment and in the lab experiment they negotiate on elements of a product bundle. The experiments control the course of alternating offer multi-issue negotiations and use between-subject comparisons of preferences after the negotiations are finished. Data on subjects' choices after having negotiated, ratings of satisfaction with the negotiated agreement, ratings of the complexity of choices among different agreements, and response times are used to test hypotheses on (1) the subjects' willingness to accept a worsening on an issue in the agreement, (2) their preferred product, (3) the sources of utility, and (4) the subjects' preference uncertainty. All of these measures are in line with predictions by the attachment effect model whereas most of them allow rejecting the applicability of the rational choice model. The only statistic that does not allow to reject either model concerns consumption utility and both models coincide in the respective prediction. The data does, however, not allow to differentiate between the attachment effect and the quasi-endowment effect, i.e. the endowment effect for an objectively un-owned item (Heyman, Orhun, and Ariely, 2004). The interpretation of the observed behavior via expectations in future endowment appears, however, more intuitive than via misperception of property rights. Thus, overall it is concluded that the experiments favor the presence of the attachment effect in negotiations.

The most important empirical results are as follows:

- Negotiators evaluate agreements in multi-issue negotiations not only in absolute terms but additionally as issue-wise gains or losses relative to a reference point that is formed during the negotiation.
- Frequently offering agreements that are generous on one issue can significantly increase the counterparty's valuation for this issue.
- Offers can change a negotiator's trade-offs between issues. Issues on which offers are generous become more important relative to issues on which offers are ungenerous.
- Negotiators obtaining an unexpectedly good agreement tend to be more satisfied with the result than negotiators getting an unexpectedly bad agreement irrespective of the absolute terms of the agreement.
- The attachment effect can influence negotiators' preference uncertainty. Negotiators choosing an object they became attached to while negotiating are less uncertain about their choice than negotiators that choose an object they are not attached to.
- Individual parameter estimates and the overall model fit show that the attachment effect model organizes data significantly better than a rational choice model. At the beginning of a negotiation, subjects are rather optimistic in the agreement they might achieve. Subsequently, reference points gradually move towards the value offered by either party. A negotiator's own offer thereby can lead to a strong decrease of her reference point. It does, however, not increase her reference point in case the offer is higher than the reference point. Furthermore, a negotiator's reference point can be influenced by her

counterparty's offer: Increases of the reference point thereby seem to be weaker than decreases.

Overall, both experiments provide suggestive evidence in favor of the attachment effect model. Thus, the theoretically derived attachment effect really is present in alternating offer negotiations (at least under the conditions employed in the experiments). Furthermore, the effect is substantial as, in the internet experiment, subjects' valuation of a feature of the apartment to be rented differed by a factor of four.

The parameter estimation has external validity. It is applied to data from the internet experiment—which was not used in the estimation— and allows to rank subjects with respect to their reference points as predicted theoretically on basis of the offers exchanged. This ranking is significantly correlated with the ranking of subjects with respect to their preferences as they were revealed via the willingness to accept a worsening in an attribute of the tenancy contract. Thus, the estimation produced a reasonable and reliable quantification of the attachment effect and allows predicting the effect of single offers on reference points.

Methodology

Besides advancing the understanding of preferences and behavior in negotiations, the work adds to the methodology of negotiation and market engineering (Weinhardt, Holtmann, and Neumann, 2003; Kersten, 2003; Ströbel, 2003; Weinhardt, Neumann, and Holtmann, 2006; Weinhardt and Gimpel, 2006) by showing how the interplay of field and lab experiments can increase validity and robustness of results. Both experiments are self-sufficient and independently constitute evidence for the attachment effect. In an engineering context when economic institutions and systems are designed and tested for their implementation in the field, it will usually be impossible to determine the exact conditions under which agents will participate. In such cases it is essential to obtain robust results that are not highly dependent on details of a theoretical model or an experimental design. This can be achieved by series of experiments on the one hand and by abandoning the abstraction of traditional, context-free economic laboratory experiments on the other hand. What on first sight seems to be the less controlled environment of a field experiment can increase robustness and transferability of results.

Currently, there is a trend towards the usage of field rather than traditional lab experiments. Most of these field experiments are not meant to replace lab experiments but to supplement them as each kind of experiment has its own strength and weaknesses. Lab experiments allow detailed control of objective elements in the subjects' environment like, for example, the information structure or the level of distraction during the experiment. On the other hand, abstract induced-value settings with instructions that show hardly any relation to real life economic situations might result in subjects bringing their own, uncontrolled, and unobserved field referents to the lab (Harrison and List, 2004) and might be inappropriate in a (negotiation and market) engineering context. To combine the strength of different types of experiments, the two experiments reported in the present work differ with respect to several aspects like the object of negotiation, the stakes, the nature of the environment, and the information structure.

A high-level comparison of the two experiments shows that both provide about the same conclusion: there is an attachment effect in multi-issue negotiations. Effort and costs, on the other hand, differ sharply. While the sample size in the lab was just about twice as large as over the internet, direct costs for the subjects' incentives were about 25 times higher in the lab. Furthermore, recruiting subjects, preparing the experiment, and conducting the sessions took a multiple of the time for preparing and conducting the internet experiment. Finally, the lab experiment had higher requirements concerning the infrastructure: an experimental lab was necessary whereas a regular personal computer as server and internet access were sufficient for the internet experiment. Overall, the lab experiment is closer to traditional standards in experimental economics and the combination of the two experiments certainly strengthens the result. The close conformance of results suggests, however, that the usage of internet experiments (potentially instead of pilot experiments commonly run in the lab prior to the replication of a lab experiment) may be beneficial prior to conducting a lab experiment.

Related Work

The present study relates to research on negotiation analysis in general and, more specifically, to common biases in negotiations (Raiffa, 1982; Neale and Bazerman, 1991; Bazerman and Neale, 1992; Raiffa, 2003; Bazerman, 2006). Negotiators oftentimes fail to reach mutually beneficial agreements. To help them in negotiating better agreements, researchers in negotiation analysis take an asymmetric prescriptive/descriptive approach (Raiffa, 1982). The descriptive part aims at understanding common patterns in the cognition and behavior of negotiators who are not fully rational in a game theoretic sense. Based on this descriptive foundation, the prescriptive part of negotiation analysis advises (individual) negotiators how to negotiate rationally given their own cognitive constraints and the likely cognition and behavior of their counterparty (e.g. Bazerman and Neale, 1992; Raiffa, 2003). In this context, the study of reference-dependent preferences in multi-issue choice is most closely related to the work by Tversky and Kahneman (1991) and other studies of prospect theory. It is inspired by empirical evidence for the endowment effect (e.g. Kahneman, Knetsch, and Thaler, 1990; Strahilevitz and Loewenstein, 1998), auction fever (e.g. Heyman, Orhun, and Ariely, 2004; Ku, Malhotra, and Murnighan, 2005), and theoretical modeling of the effect expectations can have on preferences (Köszegi and Rabin, 2006).

Reference-dependent preferences and endogenous reference points in single-issue negotiations have been studied theoretically and empirically by some authors during the last years: Shalev (2002) and Compte and Jehiel (2006), for example, present equilibrium models for bargaining games with endogenous reference points and Kristensen and Gärling (1997a) as well as Curhan, Neale, and Ross (2004) find experimental evidence for the change of preferences in negotiations. With respect to multi-issue negotiations, the most closely related work—which is the basis for parts of Chapter 4—is presented by Gimpel (2007). Furthermore, the present work relates to research on engineering negotiations and other market mechanisms (Weinhardt, Holtmann, and Neumann, 2003; Kersten, 2003; Weinhardt, Neumann, and Holtmann, 2006; Weinhardt and Gimpel, 2006).

Research Questions

Based upon the analysis of the attachment effect, the four questions posed in the introduction can be answered as follows:

1. *Are preferences endogenous to negotiations, i.e. are they influenced by the specific course of a negotiation?*
   Yes, they are. Obviously it cannot be assured that this is the case for every negotiator and every single negotiation but there are negotiations in which preferences are influenced by the course of the negotiation. Most prominently this can be expected if a negotiator is relatively inexperienced with the object of bargaining and when she does not explicitly assess her preferences and trade-offs.
2. *Can models that allow for endogeneity of preferences predict behavior significantly better than models relying on exogenous preferences?*
   Yes, they can. The attachment effect model proposed in Chapter 3 organizes data from the two experiments significantly better than a rational choice model. The increased complexity of the model is an obvious downside for some potential applications. However, as

the game theoretic equilibrium analysis of bilateral multi-issue negotiations under incomplete information is problematic anyways, one might not loose much tractability by accounting for the attachment effect. On the other hand one gains increased congruence with reality.

3. *Is there a systematic bias of preferences depending on the offers exchanged in a negotiation?*
   Yes, there is. Preferences are systematically affected by offers via the attachment effect.

4. *If it is the case that preferences are reference-dependent: How is the reference point determined?*
   Offers influence a negotiator's expectations which in turn might shift her reference point. Thus, the evaluation of outcomes as gains or losses on single issues and the negotiator's preferences might change. More formally, this causal relationship was modeled via a recursive definition of time-dependent reference points.

The attachment effect is a systematic bias that affects negotiators—this descriptive result can be used for the prescriptive part of negotiation analysis. The attachment effect classifies as a common bias in negotiations and joins a set of established biases like anchoring and adjustment, framing, the availability bias, overconfidence, the fixed pie illusion, the illusion of conflict, reactive devaluations, escalation of conflict, ignorance of the other's behavior, and egocentrism (see Sec. 3.1.3 for a description of these biases).

Knowledge of the attachment effect can be used for creating or destroying gains from trade, for debiasing oneself, for systematically influencing the counterparty's preferences, and for engineering electronic negotiation support systems. These implications are discussed in more detail in the following.

6.1.2 Implications

Once the existence of the attachment effect is assumed for a negotiation, the bias has several implications: On a collective, global perspective gains from trade between two parties can either be created or destroyed; on the individual level, negotiators might either be concerned with falling prey to the bias themselves or with the question how they can influence the counterparty; finally, researchers and software engineers might acknowledge the attachment effect when designing negotiation support systems as will be outlined below. Furthermore, Kahneman (1992) points out further implications of static, exogenous reference points in negotiation, e.g. the aversion to make concessions.

Collective Perspective

A third party observing a bilateral negotiation might observe that the two negotiators jointly search the agreement space for a mutually acceptable outcome.[1] However, by searching and exchanging offers, the parties' evaluation of the different possible agreements changes. Thus, it can happen that after negotiating for some time there is no agreement left that both parties would prefer to their best alternative in case the negotiation fails (i.e. their BATNA; cf. Sec. 3.3.4). On the contrary, gains from trade can emerge where they did not exist at the beginning of the negotiation. The negotiators' expectations, reference points, and preferences can change in a way that one or several agreements become mutually beneficial compared to the respective outside option. Thus they become acceptable and an agreement can be reached.

From a collective perspective, the detrimental effect attachment can have on the possibility of bilateral trade is devastating as it reduces welfare. Judging on the desirability of the creation of gains from trade is not so easy. On the one hand, it could be argued that at the time the parties agree, both are satisfied with the agreement and perceive it as better than their respective outside options. Thus, the agreement is—at the time it is reached—Pareto superior to not agreeing at all and hence increases welfare. On the other hand, it can be argued that the increased attractiveness is 'irrational' and the parties agree on an outcome based on a spontaneous, inconsiderate evaluation. Which of the two perspectives is taken is a matter of philosophy and not further discussed here.

Individual Perspective

In a negotiation analysis context, the results of the present study can be used for prescriptive advice to negotiators: either for debiasing or to systematically affect the counterparty's attitude towards the object of negotiation. An individual negotiator who becomes aware of the attachment effect might be concerned with not falling prey to it. The negotiator might judge the attachment effect as irrational and might want to avoid a change of her preferences. The process of consciously avoiding a bias is termed debiasing. Debiasing is a common advice in the negotiation analysis literature. Once a negotiator is aware of the attachment bias, she can aim at avoiding it by, for example, externalizing her preferences prior to the negotiation. This means that she considers

---

[1] Such a third party might be a human, an organization, or a negotiation support system.

possible agreements and consciously assesses her preferences. At best, the negotiator writes them down or communicates her preferences to someone else (e.g. a negotiation support system) who can remind her of what she wanted to achieve prior to the negotiation. Assessing ones own preferences prior to a negotiation is, e.g., proposed by Raiffa (1982, esp. Ch. 9) and Bazerman and Neale (1992, Ch. 9) as one fundamental requirement for negotiating rationally. Curhan, Neale, and Ross (2004) present evidence on endogenous preference changes in negotiations and attribute it to dissonance and reaction theory (cf. Sec. 1.1 and 3.3.2). They report that when they explained their hypotheses, 'participants for the most part doubted that their expressed preferences had changed from one round to the next as a result of the negotiation process' (p. 145). Furthermore, elicitation of preferences prior to negotiating reduced the endogenous changes of preferences. The fact that changes seem to be unconscious and that they are affected by assessing and externalizing preferences both favor the supposition that a debiasing is possible.

Manipulation of the counterparty's preferences—or, in a more neutral wording, a systematic influence on her preferences—can be achieved by a conscious selection of offers that direct the reference point in a desired direction. A requirement is that the counterparty is prone to the attachment effect for the object of negotiation. Such a manipulation can serve a negotiator in either of two ways: Firstly, it can result in an agreement that the negotiator wants to achieve but that would not have been accepted by the counterparty without attachment. Secondly, and more benevolently, the negotiator might worry about unfavorable reference points and construction of preferences and might try to avoid the case that gains from trade are destroyed during the negotiation.

Electronic Negotiation Support

A final implication of the attachment effect concerns researchers and software engineers building electronic NSS (negotiation support systems). Such systems offer different degrees of support and help in, for example, the communication process by simply transmitting offers and keeping record of previous offers. Furthermore, analytical support can assist in evaluating offers in case the NSS elicited the negotiator's preferences. Further steps of negotiation support are assistance in offer generation, in arbitration, or in searching for post-negotiation improvements. See e.g. Kersten (2004) for an overview on negotiation support systems in general. Specific NSS and their features are, for example, presented by Schoop, Köhne, and Staskiewicz (2004) with respect to the

.communication support offered by the Negoisst system, Thiessen and Soberg (2003) concerning the analytical support of post-negotiation phases in the Smartsettle system, and Kersten and Noronha (1999) for the analytical and communication support the Inspire system offers.

For the analytical support offered by NSS, the implications of the attachment effect are as follows (Gimpel, 2007): (1) Systems might warn their users to avoid the sketched bias and, thus, help in debiasing. Furthermore, eliciting a utility function, automatically evaluating offers, and reminding the negotiator about her ex-ante preferences (or at least showing the discrepancy of current preferences to ex-ante preferences) can likely diminish the bias. (2) If the system deals with post-negotiation improvements, it should re-elicit the negotiators' preferences at the end of a negotiation. As preferences might have changed during the negotiation this is necessary to assure that a proposed improvement really is perceived as beneficial by the negotiators. (3) If the changed preference structure is temporary and the (true) ex-ante preferences recur—which was not tested in the present work—, the system should propose improved agreements with considerable delay after the end of the negotiation. (4) Furthermore, a NSS could assist its user in the offer generation process and might recommend offers counteracting the attachment effect in the counterparty's cognition and evaluation of offers. This might be achieved by, e.g., making pairs of offers instead of single offers: If the two offers are sufficiently different on each issue individually, the counterparty's expectations likely are not clearly oriented to a specific agreement and, thus, the negotiator might be less prone to the attachment bias. The NSS could assure that both offers give approximately the same utility for the negotiator proposing them. Testing this suggestion and the efficiency of the other implications for NSS is, however, up to future work.

To date there is no NSS that explicitly addresses these points. Some do, however, allow adjusting the utility function when a negotiator finds the initially specified function (e.g. the Inspire system; Kersten and Noronha, 1999).

6.2 Limitations of the Present Work

The present work studies the attachment effect in bilateral alternating offer multi-issue negotiations between monolithic parties—thereby the scope of the work is self-evidently limited: It does neither consider multilateral negotiations, nor static or single-sided ones; it does not address single-issue bargaining and the study of non-monolithic parties is beyond the limits of the present work. Furthermore, it builds on experimentation and the usual concerns about inductive reasoning in science apply.

The question of external validity probably is the most common line of attack against experimental economics and any other experimental science. Can results obtained in a relatively simple and abstract experiment be generalized to more complex 'real world situations'? To define external validity, it is convenient to differentiate it from internal validity (Guala, 2005, Ch. 7): Internal validity is achieved when a causal relationship is established in a laboratory setting. Given an experimental environment E, a factor O (or a set of factors), and an effect P, internal validity means that the experimenter correctly conjectures a relationship $E : O \rightarrowtail P$, i.e. given E, O causes P, and O is really the cause of P. The experiment is externally valid if O causes P not only in environment E, but also in F, G, etc. See Guala (2005, Ch. 7 and 8–11) for an extensive methodological discussion of external validity in experimental economics. External validity bases on the general principle of induction that Smith (1982a) terms *parallelism* of economic experiments; he formulates it as follows:

> 'Propositions about the behavior of individuals and the performance of institutions that have been tested in laboratory microeconomies apply also to nonlaboratory microeconomies where similar *ceteris paribus* conditions hold.' (p. 936; italics in the original.)

The transferability of experimental results on the attachment effect to the field thus depends on the question which ceteris paribus conditions are 'similar' and which are not. Unfortunately, there is no unanimous way to assess the similarity of two environments E and F and, thus, no general model that would allow to follow $F : O \rightarrowtail P$ from observing $E : O \rightarrowtail P$ in an experiment.

The experimental control, especially the control over offer sequences by the first landlord and agent A, respectively, strengthens internal validity and allows to claim that the observed treatment difference with respect to ex-post revealed preferences, the sources of utility, and preference uncertainty are caused by the agents' offer sequences. This conclusion is a statistical conclusion—it bases on the test of hypotheses and it might be false, although this is unlikely.

Whether the experiments have any external validity beyond the specific environments studied cannot be proven based on the available data. It could just be tested by further empirical studies in other environments. The two arguments that favor external validity to some degree are, firstly, that the internet and the lab experiment differ with respect to several design features (commodity, nature of the stakes, nature of the environment, incentive compatibility, preference elicitation method, and time when the experiment was conducted). Despite these differences, results from both experiments suggest the same effect of offers (O) on preferences (P). Thus, in the notation introduced above, $E : O \rightarrowtail P$ and $F : O \rightarrowtail P$ have been shown, where E is the environment in the internet experiment, F the environment in the lab experiment.

Secondly, empirical evidence gathered with the negotiation support system Inspire shows a discrepancy between the negotiators' behavior and their reported utility functions (cf. Ch. 1; Vetschera, 2004b; Block et al., 2006). This difference might be explained by a change in the negotiators' preferences during the negotiation. However, due to the less controlled setting in Inspire negotiations, it is not possible to directly assess this and bring up an internally valid conclusion about reference points based on Inspire data (Block et al., 2006).

The most relevant limitations for transferability of the results with respect to the negotiations studied in the present work are as follows:

Alternating offer: Throughout the work it was assumed that parties take turns in proposing offers. In the field, other negotiation protocols occur frequently. It seems likely that the attachment effect would as well be present in a single-sided negotiation, i.e. a negotiation in which just one party makes offers and the other party solely rejects or accepts these offers. In a static, sealed-bid negotiation, on the other hand, the results might no longer hold as there are no intermediate outcomes that might affect expectations and attachment.

Multi-issue: As single-issue negotiations are a special case of multi-issue negotiations, the attachment effect model transfers to single-issue cases. Whether the negotiators' perception and hence the experimental results would be comparable can, however, not be determined based on the present study, even if the experiment by Kristensen and Gärling (1997a) suggests an attachment effect in single-issue negotiations.

Monolithic parties: The attachment effect assumes that the individual cognition and mental processing of offers is biased. It is questionable whether the effect transfers to non-individual decision-making of non-monolithic parties. Kahneman (1992), for example, expects that reference points affect negotiations even with non-monolithic parties.

Single bilateral negotiation: The present work cannot assure whether the attachment effect applies in multilateral negotiations although there is no obvious reason why it should not apply. Furthermore, the effect of multiple parallel negotiations with different counterparties and the presence of third parties like mediators have not been addressed.

Agenda: In the negotiations studied in the present work, all issues are negotiated simultaneously. A sequential negotiation of issues would likely increase the attachment effect for the issues that are settled first. This has, however, not been addressed in the experiments.

Real world economic environments are highly complex and any experiment can just resemble a simplification. The two experimental designs differ with respect to several factors and thereby increase transferability of the results. Other design features are, however, constant and thus constitute limitations for transferability. Most prominently these are the following:

Automated counterparty: In both experiments, subjects negotiated with software agents. Whether the same results would hold in the interaction between two human negotiators cannot be answered based on the present work although the role of emotions in decision-making suggests that this might be the case (cf. Sec. 4.1.4): The play against a computer likely is less emotion-laden than interacting with another human and, thus, can be expected to reduce the attachment effect. Hence, the effect might be even more pronounced in non-automated negotiations.

Message space: The experimental system limited messages to offering points in the agreement space (thereby rejecting the previous offer if any) or accepting an offer. Withdrawing offers and exchanging arguments or other free text or non-verbal communication was impossible. Whether an increase of the message space has an influence on the attachment effect is beyond the scope of the present work.

Subject pool: Both experiments used a student subject pool as it is common in most economic experiments. The subject pool certainly is unrepresentative for the whole population in terms of age and education. Thus, transferability of results to a wider subject pool cannot be guaranteed. There is, however, evidence that most experimental results obtained with student subjects hold outside the lab as well (e.g. Friedman and Sunder, 1994, Ch. 4; Harrison and List, 2004) and other studies suggest that the behavior of students is closer to the assumption of rational choice theory than the be-

havior of the general population (e.g. Bellemare, Kröger, and van
Soest, 2005).

Experience: The objects of negotiation, i.e. the tenancy contract and
the consumer goods, were chosen in a way that subjects likely have
experience with the issues and can evaluate trade-offs. On the other
hand, it was expected that they are not 'too experienced' and do
not (yet) have well defined and invariable preferences. Assessing
the boundaries of the attachment effect with respect to negotiators'
experience was not in the scope of the study at hand.

Nature of the stakes: In the lab experiment, stakes were relatively low
and a subject's average payment evaluated by retail prices of the
goods was about € 15. In the internet experiment (hypothetical)
stakes were much higher. Depending on the time horizon of subjects' planning—which was neither controlled nor monitored—the
stakes range from several hundred to several thousand euros.[2] Thus,
the attachment effect occurs at quite different stakes. Whether the
same results would be obtained in negotiations about real outcomes
with a value of several thousand, several hundred thousand, several
million, etc. euros is, however, beyond the limits of the present work.

The question whether the same effect of offers on preferences ($O \rightarrowtail$
P) transfers to other environments than E (the internet) and F (the
lab), is subject to speculation. If the other environment is an environment E' or F' that just slightly differs from either setting, this appears
very likely. A slight change might, for example, be having green coffee
mugs instead of blue ones, allowing up to 20 offers instead of just 12, or
negotiating about 5 issue rather than 3. For more fundamental changes,
this cannot be said—e.g. when the stakes are raised to a million euros,
when issues are complements or perfect substitutes, when the counterparty is a human instead of a computer, or the negotiator is a trained
professional negotiator rather than a student. There is no general way
to assess where exactly the 'similar ceteris paribus' conditions end.

Formalization of the Attachment Effect

A final limitation of the present study is the formalization of the attachment effect in Section 3.3.2. There it was said that the 'attachment
effect [...] *can* be modeled as follows' (p. 101; italics added). It could,
however be done differently as well. Including the entire history of offers

---

[2] On the other hand, it could be argued that the stakes in the internet experiment
were zero, as the tenancy contracts were hypothetical. However, the subjects'
behavior and especially differences in behavior depending on the agent's offer
sequence suggest that they sincerely considered the contracts.

instead of just the current offer would, for example, be one extension and having time dependent update functions would be another. Ignoring the effect of a negotiator's own offers might be a simplification, etc.

The specific functional relationships assumed are inspired by empirical findings on the endowment effect (e.g. Strahilevitz and Loewenstein, 1998) and theoretical models on endogenous preferences in single-issue bargaining (c.g. Compte and Jehiel, 2006). Furthermore, the success of the chosen model in organizing the data from the two experiments with relatively simple linear update functions proofs that this formalization is reasonable and powerful, even if it might not be the only possible one.

6.3 Future Work

The previous discussion of limitations of the present work pointed out the potential for extending the study of the attachment effect— or more generally the endogenous nature of preferences in markets—in various directions. This final section takes a view beyond the results presented so far and sketches the most interesting and relevant directions for future work. They deal with transferability of the results, testing the implications of the attachment effect, advancing the methodology of experimental economics, and searching for the boundaries of the attachment effect in other negotiation and market processes.

External Validity

As discussed in the previous section, a single experiment can never show that the same effect would occur again or in any other environment. A pair of experiments is better but again the question of transferability occurs. Future work to test the applicability of the attachment effect in bilateral alternating offer negotiations might gather further empirical evidence to corroborate and extend the results found in the present study. Specifically, experimentation with (1) negotiations among two humans, with (2) experienced subjects or goods for which subjects have more experience, with (3) other types of goods and services, and with (4) non-monolithic parties would be of great interest. Furthermore, field rather than experimental data could strengthen evidence for the attachment effect. Here it will be a challenge to find non-experimental negotiations in which the negotiators' preferences can be assessed in enough detail to detect an attachment effect.

Test Implications

The two experiments presented in the present work only test the most direct implication of the attachment effect: they test whether preferences and, thus, choices depend on the history of offers exchanged. Future work might include assessing (1) in how far debiasing via, for example, explicit warnings is possible and (2) whether the case that gains from trade that initially exist are destroyed by negotiating is relevant in non-experimental negotiations. Furthermore, implications for engineering negotiation support systems were briefly mentioned but not yet tested empirically.

Methodology of Experimental Economics

Both experiment designs employed in the present study rely on software agents as counterparties for human negotiators. Furthermore, numerous other experiments used software agents before. The predominant reason is that software agents taking the role of some players allow the researcher a great deal of control over the experiment. A nice side effect is that oftentimes the number of independent observations is increased and, thus, experimental costs are reduced. Nevertheless, up to date only few studies addressed the question how humans perceive the play against computers and how the nature of the counterparty influences behavior (cf. Sec. 4.1.4). Further investigating this question would on the one hand be a contribution to the methodology of experimental economics and, on the other hand, it would have relevance for the interactions of humans and automated decision-makers in many real life situations that occur increasingly often.

Negotiation and Market Engineering

Finally, testing the presence of the attachment effect and searching for its boundaries in other negotiation protocols and market mechanisms is up to future work. Köszegi and Rabin (2006) discuss the role of expectations and attachment in posted price markets, for example. Recent experimental evidence suggests the existence of an attachment effect in iterative single-unit auctions and the present study investigates the attachment effect in multi-issue alternating offer negotiations. These market mechanisms and interaction protocols are undoubtedly relevant to many real world contexts. There are, however, numerous other economic interactions for which the attachment effect might be relevant, e.g. repeated sealed-bid (static) negotiations, negotiations with only one side proposing agreements, multi-unit auctions, and structured double-sided exchanges like financial markets. Predicting theoretically and testing empirically where the attachment effect might apply in these non-individual decision-making situations would be a further contribution to the field of negotiation and market engineering.

References

Abele, S., K.-M. Ehrhart, and M. Ott (2006). An experiment on auction fever. Working Paper.

Agresti, A. (1990). *Categorical Data Analysis*. New York: John Wiley & Sons.

Allingham, M. (2002). *Choice Theory*. Oxford: Oxford University Press.

Ariely, D., G. F. Loewenstein, and D. Prelec (2003). Coherent arbitrariness: Stable demand curves without stable preferences. *The Quarterly Journal of Economics 118*(1), 73–105.

Ariely, D., G. F. Loewenstein, and D. Prelec (2006). Tom sawyer and the construction of value. *Journal of Economic Behavior and Organization 60*(1), 1–10.

Ariely, D. and I. Simonson (2003). Buying, bidding, playing, or competing? value assessment and decision dynamics in online auction. *Journal of Consumer Psychology 13*(1&2), 113–123.

Arlen, J., M. Spitzer, and E. Talley (2002). Endowment effects within corporate agency relationships. *Journal of Legal Studies 31*(1), 1–37.

Arrow, K. J. (1982). Risk perception in psychology and economics. *Economic Inquiry 20*(1), 1–9.

Arrow, K. J. (1986). Rationality of self and others in an economic system. *The Journal of Business 59*(4, Part 2), S385–S399.

Ausubel, L. M., P. Cramton, and R. J. Deneckere (2002). Bargaining with incomplete information. In R. Aumann and S. Hart (Eds.), *Handbook of Game Theory With Economic Applications Volume 3*, pp. 1897–1946. Amsterdam: North-Holland.

Ausubel, L. M. and R. J. Deneckere (1992). Durable goods monopoly with incomplete information. *Review of Economic Studies 59*, 795–812.

Babcock, L., G. F. Loewenstein, S. Issacharoff, and C. F. Camerer (1995). Biased judegments of fairness in bargaining. *American Economic Review 85*(5), 1337–1343.

Barnard, G. A. (1945). A new test for 2×2 tables. *Nature 156*, 177.

Barnard, G. A. (1947). Significance tests for 2×2 tables. *Biometrika 34*, 123–138.

Barnard, G. A. (1949). Statistical inference (with discussion). *Journal of the Royal Statistical Society. Series B 11*, 115–149.

Barnard, G. A. (1984). Discussion of 'tests of significance in 2×2 tables' by f. yates. *Journal of the Royal Statistical Society. Series A 147*, 449–450.

Bateman, I., A. Munro, B. Rhodes, C. Starmer, and R. Sudgen (1997). Does part-whole bias exist? an experimental investigation. *Economic Journal 107*, 322–332.

Bazerman, M. and J. Carroll (1987). Negotiator cognition. In *Research in Organizational Behavior*, Volume 9, pp. 247–288. Greenwich: JAI Press.

Bazerman, M. H. (2006). *Judgment in Managerial Decision Making* (6. ed.). New York: John Wiley & Sons.

Bazerman, M. H., J. R. Curhan, D. A. Moore, and K. L. Valley (2000). Negotiation. *Annual Review of Psychology 51*, 279–314.

Bazerman, M. H., T. Magliozzi, and M. A. Neale (1985). Integrative bargaining in a competitive market. *Organizational Behavior and Human Decision Processes 35*, 294–313.

Bazerman, M. H. and M. A. Neale (1982). Improving negotiation effectiveness under final offer arbitration: the role of selection and training. *Journal of Applied Psychology 67*, 543–548.

Bazerman, M. H. and M. A. Neale (1983). Heuristics in negotiation: limitations to effective dispute resolution. In M. H. Bazerman and R. J. Lewicki (Eds.), *Negotiating in Organizations*. Thousand Oaks: Sage Publications.

Bazerman, M. H. and M. A. Neale (1992). *Negotiating Rationally*. New York: Free Press.

Bearden, J. N., W. Schulz-Mahlendorf, and S. Huettel (2006). An experimental study of von Neumann's two-person [0,1] poker. Working Paper.

Bellemare, C., S. Kröger, and A. van Soest (2005). Actions and beliefs: Estimating distribution-based preferences using a large scale experiment with probability questions on expectations. Working Paper.

Bern, D. J. (1967). Self-perception: An alternative interpretation of cognitive dissonance phenomena. *Psychological Review* 74(3), 183–200.

Bern, D. J. (1972). Self-perception theory. In L. Berkowitz (Ed.), *Advances in Experimental Social Psychology*, pp. 1–62. San Diego: Academic Press.

Bernoulli, D. (1954). Exposition of a new theory on the measurement of risk. *Econometrica* 22(1), 23–36. Original 1738.

Best, D. J. and D. E. Roberts (1975). Algorithm as 89: The upper tail probabilities of spearman's rho. *Applied Statistics* 24, 377–379.

Bettman, J. R., E. J. Johnson, M. F. Luce, and J. W. Payne (1993). Correlation, conflict and choice. *Journal of Experimental Psychological: Learning, Memory and Cognition* 19(4), 931–951.

Bettman, J. R., M. F. Luce, and J. W. Payne (1998). Constructive consumer choice processes. *Journal of Consumer Research* 25(3), 187–217.

Bichler, M., G. E. Kersten, and S. Strecker (2003). Towards a structured design of electronic negotiations. *Group Decision and Negotiation* 12(4), 311–335.

Binmore, K., M. J. Osborne, and A. Rubinstein (1989). Noncooperative models of bargaining. In R. Aumann and S. Hart (Eds.), *Handbook of Game Theory with Economic Applications*, Volume 1, Chapter 7, pp. 179–225. Amsterdam: North-Holland.

Bishop, R. C. and T. A. Heberlein (1979). Measuring values of extramarket goods: Are indirect measures bises? *American Journal of Agricultural Economics* 61, 926–930.

Bizman, A. and M. Hoffman (1993). Expectations, emotions, and preferred responses regarding the arab-israeli conflict: an attributional analysis. *Journal of Conflict Resolution* 37, 139–159.

Block, C., H. Gimpel, G. Kersten, and C. Weinhardt (2006). Reasons for rejecting pareto-improvements in negotitations. In S. Seifert and C. Weinhardt (Eds.), *Group Decision and Negotiation (GDN) 2006*, pp. 243–246. Karlsruhe: Universitätsverlage Karlsruhe.

Böckenholt, U., D. Albert, K. M. Aschenbrenner, and F. Schmalhofer (1991). The effect of attractiveness, dominance, and attribute differences on information acquisition in multiattribute binary choice. *Organizational Behavior and Human Decision Processes* 49, 258–281.

Bohm, P. (1972). Estimating the demand for public goods: An experiment. *European Economic Review* 3, 111–130.

Bolton, G. E. and A. Ockenfels (2000). Erc: A theory of equity, reciprocity, and competition. *American Economic Review 90*(1), 166–193.

Bottom, W. and A. Studt (1993). Framing effects and the distributive aspect of integrative bargaining. *Organizational Behavior and Human Decision Processes 56*, 459–474.

Bowles, S. (2004). *Microeconomics: Behavior, Institutions, and Evolution.* Princeton: Princeton University Press.

Bowles, S., J.-K. Choi, and A. Hopfensitz (2003). The coevolution of individual behaviors and group level institutions. *Journal of Theoretical Biology 223*(2), 135–147.

Brehm, J. W. (1966). *A Theory of Psychological Reactance.* San Diego: Academic Press.

Brookshire, D. S. and D. L. Coursey (1987). Measuring the value of a public good: An empirical comparison of elicitiona procedures. *American Economic Review 77*(4), 554–566.

Brookshire, D. S., A. Randall, and J. R. Stoll (1980). Valuing increments and decrements in natural resource service flows. *American Journal of Agricultural Economics 62*, 478–488.

Bush, P. D. (1993). The methodology of institutional economics: A pragmatic instrumentalist perspective. In M. R. Tool (Ed.), *Institutional Economics: Theory, Method, Policy.* Dordrecht: Kluwer Academic Publishers.

Camerer, C. and G. F. Loewenstein (1993). Information, fairness, and efficiency in bargaining. In B. Mellers and J. Baron (Eds.), *Psychological Perspectives on Justice: Theory and Applications*, pp. 155–181. New York: Cambridge University Press.

Camerer, C. F. (1995). Individual decision making. In J. H. Kagel and A. E. Roth (Eds.), *Handbook of Experimental Economics*, pp. 587–703. Princeton: Princeton University Press.

Camerer, C. F. (2001). Prospect theory in the wild: Evidence from the field. In D. Kahneman and A. Tversky (Eds.), *Choices, Values, and Frames*, pp. 288–300. Cambridge: Cambridge University Press.

Camerer, C. F., G. F. Loewenstein, and D. Prelec (2005). Neuroeconomics. *Journal of Economic Literature 43*(1), 9–64.

Campbell, M., A. J. Hoane, and F.-H. Hsu (2002). Deep blue. *Artificial Intelligence 134*, 57–83.

Carroll, J., M. Bazerman, and R. Maury (1988). Negotiator cognitions: a descriptive approach to negotiators' understanding of their opponents. *Organizational Behavior and Human Decision Processes 41*, 352–370.

Carter, J. (1982). *Keeping Faith*. London: William Collins Sons & Co Ltd.

Carter, R. (1998). *Mapping the Mind*. London: Weidenfeld & Nicholson.

Chatterjee, K. and W. Samuelson (1983). Bargaining under incomplete information. *Operations Research 31*(5), 835–851.

Coase, R. H. (1960). The problem of social cost. *Journal of Law and Economics 3*, 1–44.

Colander, D. (1994). The art of economics by the numbers. In R. E. Backhouse (Ed.), *New Directions in Economic Methodology*, pp. 35–49. Routledge.

Compte, O. and P. Jehiel (2006). Bargaining with reference dependent preferences. Working Paper.

Connolly, T., H. R. Arkes, and K. R. Hammond (1999). *Judgment and Decision Making*. Cambridge: Cambridge University Press.

Conover, W. J. (1999). *Practical Nonparametric Statistics* (3. ed.). New York: John Wiley & Sons.

Cornfield, J. (1956). A statistical problem arising from retrospective studies. In J. Negman (Ed.), *Proceeding of the Third Berkeley Symposium, Volume IV*, Berkeley, pp. 135–148. University of California Press.

Coupey, E., J. R. Irwin, and J. W. Payne (1998). Product category familarity and preference construction. *Journal of Consumer Research 24*(4), 459–468.

Cox, D. R. (1984). Discussion of 'tests of significance in 2×2 tables' by f. yates. *Journal of the Royal Statistical Society. Series A 147*, 451.

Cox, J. C. and D. M. Grether (1996). The preference reversal phenomenon: Response mode, markets and incentives. *Economic Theory 7*(3), 381–405.

Cramer, J. S. (1986). *Econometric applications of Maximum Likelihood methods*. Cambridge: Cambridge University Press.

Cubitt, R. P., C. Starmer, and R. Sudgen (2001). Discovered preferences and the experiemental evidence of violations of expected utility theory. *Journal of Economic Methodology 8*, 385–414.

Cummings, R. G., G. W. Harrison, and E. E. Rutström (1995). Homegrown values and hypothetical surveys: Is the dichotomous choice approach incentive compatible? *American Economic Review 85*(1), 260–266.

Curhan, J. R., M. A. Neale, and L. D. Ross (2004). Dynamic valuation: Preference changes in the context of face-to-face negotiation. *Journal of Experimental Social Psychology 40*, 142–151.

D'Agostino, R. B., W. Chase, and A. Belanger (1988). The appropriateness of some common procedures for testing equality of two independent binomial proportions. *The American Statistician 42*, 198–202.

Davis, D. D. and C. A. Holt (1992). *Experimental Economics*. Princeton: Princeton University Press.

De Dreu, C. and C. McCusker (1997). Gain-loss frames and cooperation in two-person social dilemmas: a transformational analysis. *Journal of Personality and Social Psychology 72*(5), 1093–1106.

Diekmann, K., A. Tenbrunsel, P. Shah, H. Schroth, and M. Bazerman (1996). The descriptive and prescriptive use of previous purchase price in negotiations. *Organizational Behavior and Human Decision Processes 66*, 179–191.

Duersch, P., A. Kolb, J. Oechssler, and B. C. Schipper (2005). Rage against the machines: How subjects learn to play against computers. Working Paper.

Eckel, C. C. and P. J. Grossman (1996). Altruism in anonymous dictator games. *Games and Economic Behavior 16*, 181–191.

Edwards, W. (1977). How to use mutliattribute utility measurement for social decision-making. *IEEE Transactions on Systems, Man and Cybernetics 7*, 326–340.

Edwards, W. and F. H. Barron (1994). Smarts and smarter: Improved simple methods for multiattribute utility measurement. *Organizational Behavior and Human Decision Processes 60*, 306–325.

Fehr, E. and K. Schmidt (1999). A theory of fairness, competition, and cooperation. *The Quarterly Journal of Economics 114*(3), 817–868.

Fehr, e. and J.-R. Tyran (2005). Individual irrationality and aggregate outcomes. *Journal of Economic Perspectives 19*(4), 43–66.

Festinger, L. (1957). *A Theory of Cognitive Dissonance*. Stanford: Stanford University Press.

Festinger, L. and E. Aronsons (1960). The arousal and reduction of dissonance in social contexts. In D. Cartwright and A. Zander (Eds.), *Group Dynamics*, pp. 214–231. Evanston: Row Peterson.

Fiorillo, C. D., P. N. Tobler, and W. Schultz (2003). Discrete coding of reward probability and uncertainty by dopamine neurons. *Science 299*(5614), 1898–1902.

Fischer, G. W., J. Jia, and M. F. Luce (2000). Attribute conflict and preference uncertainty: The randmau model. *Management Science 46*(5), 669–684.

Fischer, G. W., M. F. Luce, and J. Jia (2000). Attribute conflict and preference uncertainty: Time and error. *Management Science 46*(1), 88–103.

Fischhoff, B. (1982). Latitudes and platitudes: How much credit do people deserve? In G. Ungson and D. Braunstein (Eds.), *New Directions in Decision Making*. New York: Kent.

Fishburn, P. (1970). *Utility Theory for Decision Making*. New York: John Wiley & Sons.

Fisher, G. W., J. Jia, and M. F. Luce (2000). Attribute conflict and preference unvertainty: The randmau model. *Management Science 46*(5), 669–684.

Fisher, G. W., M. F. Luce, and J. Jia (2000). Attribute conflict and preference unvertainty: Effects on judgement time and error. *Management Science 46*(1), 88–103.

Fisher, R. and W. L. Ury (1983). *Getting to Yes: Negotiating Agreement Without Giving In*. New York: Penguin Books.

Fisher, R. A. (1921). On the mathematical foundations of theoretical statistics. *Philosophical Transactions of the Royal Society of London Series A 222*, 309–368.

Fisher, R. A. (1925). Theory of statistical estimation. *Proceedings of the Cambridge Philosophical Society 22*, 700–725.

Fisher, R. A. (1934). *Statistical Methods for Research Workers* (5. ed.). Edinburgh: Oliver and Boyd.

Fisher, R. A. (1935a). *Design of Experiments*. Edinburgh: Oliver and Boyd.

Fisher, R. A. (1935b). The logic of inductive inference. *Journal of the Royal Statistical Society: Series A 98*, 39–54.

Fox, J. (1972). The learning of strategies in a simple, two-person zero-sum game without saddlepoint. *Behavioral Science 17*, 300–308.

Franciosi, R., P. Kujal, R. Michelitsch, V. L. Smith, and G. Deng (1996). Experimental tests of the endowment effect. *Journal of Economic Behavior and Organization 30*(2), 213–226.

Frederick, S., G. F. Loewenstein, and T. O'Donoghue (2002). Time discounting and time preference: A critical review. *Journal of Economic Literature 40*(2), 351–401.

Friedman, D. and S. Sunder (1994). *Experimental Methods: A Primer for Economists*. Cambridge: Cambridge University Press.

Fudenberg, D., D. K. Levine, and J. Tirole (1985). Infinite-horizon models of bargaining with one-sided incomplete information. In A. E. Roth (Ed.), *Game-Theoretic Models of Bargaining*, pp. 73–89. Cambridge: Cambridge University Press.

Fudenberg, D. and J. Tirole (1983). Sequential bargaining with incomplete information. *Review of Economic Studies 50*(2), 221–247.

Fudenberg, D. and J. Tirole (1991). *Game Theory*. Cambridge: MIT Press.

Fukuno, M. and K. Ohbuchi (1997). Cognitive biases in negotiation: the determinants of fixed-pie assumption and fairness bias. *Japanese Journal of Social Psychology 13*(1), 43–52.

Gilovich, T., D. Griffin, and D. Kahneman (2002). *Heuristics and Biases: The Psychology of Intuitive Judgment*. Cambridge: Cambridge University Press.

Gimpel, H. (2005). Reference-dependent preferences in multi-issue bargaining. In D. Lehmann, R. Müller, and T. Sandholm (Eds.), *Computing and Markets*, Number 05011 in Dagstuhl Seminar Proceedings. Internationales Begegnungs- und Forschungszentrum (IBFI), Schloss Dagstuhl, Germany.

Gimpel, H. (2006). Negotiation fever: Loss-aversion in multi-issue negotiations. In N. Jennings, G. Kersten, A. Ockenfels, and C. Weinhardt (Eds.), *Negotiation and Market Engineering*, Number 06461 in Dagstuhl Seminar Proceedings. Internationales Begegnungs- und Forschungszentrum (IBFI), Schloss Dagstuhl, Germany.

Gimpel, H. (2007). Loss aversion and reference-dependent preferences in multi-attribute negotiations. *Group Decision and Negotiation* (forthcoming).

Gimpel, H., H. Ludwig, A. Dan, and B. Kearney (2003). Panda: Specifying policies for automated negotiations of service contracts. In M. E. Orlowska, S. Weerawarana, M. P. Papazoglou, and J. Yang (Eds.), *Service-Oriented Computing – ICSOC 2003*, pp. 287–302. Berlin: Springer.

Gimpel, H. and J. Mäkiö (2006). Towards multi-attribute double auctions for financial markets. *Electronic Markets – The International Journal 16*(2), 130–139.

Glimcher, P. W., M. C. Dorris, and H. M. Bayer (2005). Physiological utiltiy theory and the neuroeconomics of choice. *Games and Economic Behavior 52*, 213–256.

Gneezy, U. and A. Rustichini (2000). Pay enough or don't pay at all. *Quarterly Journal of Economics 115*(3), 791–810.

Good, P. I. (2000). *Permutation tests: a practical guide to resampling methods for testing hypotheses* (2. ed.). Springer series in statistics. Berlin: Springer.

Guala, F. (2005). *The Methodology of Experimental Economics*. Cambridge: Cambridge Universtiy Press.

Gupta, S. and Z. A. Livne (1988).   Resolving a conflict situation
with a reference outcome: An axiomatic model. *Management Science 34*(11), 1303–1314.

Haaijer, R., W. Kamakura, and M. Wedel (2000). Response latencies
in the analysis of conjoint choice experiments. *Journal of Marketing Research 37*(3), 376–382.

Hardie, B. G. S., E. J. Johnson, and P. S. Fader (1993). Modeling loss
aversion and reference dependence effects on brand choice. *Management Science 12*(4), 378–394.

Harrison, G. W. and J. A. List (2004). Field experiments. *Journal of
Economic Literature 42*(4), 1013–1059.

Harrison, G. W., J. A. List, and C. Towe (forthcoming). Naturally oc-
curing markets and exogenous laboratory experiments: A case study
of risk aversion. *Econometrica*.

Hayek, F. A. (1967). The economy, science and politics. In F. A. Hayek
(Ed.), *Studies in Philosophy, Politics and Economics*, pp. 251–269.
Chicago: University of Chicago Press.

Heath, C., S. Huddart, and M. Lang (1999). Psychological factors and
stock option exercise. *Quarterly Journal of Economics 114*(2), 601–
627.

Heiner, R. A. (1985). Experimental economics: Comment. *American
Economic Review 75*(1), 260–263.

Hertwig, R. and A. Ortmann (2001). Experimental practices in economics: A methodological challenge for psychologists? *Behavioral
and Brain Sciences 24*, 383–451.

Hevner, A. R., S. T. March, J. Park, and S. Ram (2004). Design science
in information systems research. *MIS Quarterly 28*(1), 75–105.

Heyman, J. E., Y. Orhun, and D. Ariely (2004). Auction fever: The ef-
fect of opponents and quasi-endowment on product valuations. *Journal of Interactive Marketing 18*(4), 7–21.

Hoeffler, S. and D. Ariely (1999). Constructing stable preferences: A
look into dimensions of experiement and their impact on preference
stability. *Journal of Consumer Psychology 8*(0), 113–139.

Hollander, M. and D. A. Wolfe (1973). *Nonparametric statistical inference*. New York: John Wiley & Sons.

Horowitz, J. K. and K. E. McConnell (2002). A review of wta/wtp studies. *Journal of Environmental Economics and Management 44*(3),
426–447.

Hossain, T. and J. Morgan (2006). ... plus shipping and handling:
Revenue (non)equivalence in field experiments on ebay. *Advances in
Economic Analysis & Policy 6*(2), Article 3.

Houser, D. and R. Kurzban (2002). Revisiting kindness and confusion in public goods experiments. *American Economic Review 94*, 1062–1069.

Hutchinson, J. W., K. Raman, and M. K. Mantrala (1994). Finding choice alternatives in memory: Probability models of brand name recall. *Journal of Marketing Research 31*(4), 441–461.

Hyndman, K. (2005). Repeated bargaining with reference dependent preferences. Working Paper.

Irwin, J. (1935). Test of significance for differences between percentages based on small numbers. *Metron 12*(2), 83–94.

Jertila, A. and M. Schoop (2005). Electronic contracts in negotiation support systems: Challenges, design and implementation. In *Proceedings of the 7th International IEEE Conference on E-Commerce Technology (CEC 2005)*, Lost Alamitos, pp. 396–399. IEEE Computer Society.

Kagel, J. H. and A. E. Roth (Eds.) (1995). *Handbook of Experimental Economics*. Princeton: Princeton University Press.

Kahneman, D. (1992). Reference points, anchors, norms, and mixed feelings. *Organizational Behavior and Human Decision Processes 51*(2), 269–312.

Kahneman, D., J. L. Knetsch, and R. H. Thaler (1990). Experimental tests of the endowment effect and the coase theorem. *Journal of Political Economy 98*(6), 1325–1348.

Kahneman, D. and A. Tversky (1979). Prospect theory: An analysis of decision under risk. *Econometrica 47*(2), 263–292.

Kalai, E. and M. Smorodinsky (1975). Other solutions to nash's bargaining problem. *Econometrica 43*(3), 513–518.

Keeney, R. L. and H. Raiffa (1993). *Decisions with Multiple Objectives: Preferences and Value Tradeoffs*. Cambridge: Cambridge University Press.

Kenny, D. and C. Judd (1986). Consequences of violating the independence assumption in analysis of variance. *Psychological Bulletin 99*(3), 422–431.

Kenny, D. and C. Judd (1996). A general procedure for the estimation of interdependence. *Psychological Bulletin 119*(1), 138–148.

Kersten, G. E. (2003). The science and engineering of e-negotiation: An introduction. In *Proceedings of the 36th Annual Hawaii International Conference on System Sciences (HICSS'03)*, Los Alamitos. IEEE Computer Society.

Kersten, G. E. (2004). E-negotiation systems: Interaction of people and technologies to resolve conflicts. Presented at: UNESCAP Third

Annual Forum on Online Dispute Resolution Melbourne, Australia, July 2004.

Kersten, G. E. and S. J. Noronha (1999). Www-based negotiation support: design, implementation, and use. *Decision Support Systems 25*(2), 135–154.

Kiesler, S., L. Sproull, and K. Waters (1996). A prisoner's dilemma experiment on cooperation with people and human-like computers. *Journal of Personality & Social Psychology 70*, 47–65.

Kirchkamp, O. and R. Nagel (2007). Naive learning and cooperation in network experiments. *Games and Economic Behavior 58*, 269–292.

Knetsch, J. L. (1989). The endowment effect and evidence of nonreversible indifference curves. *American Economic Review 79*, 1277–1284.

Köszegi, B. and M. Rabin (2006). A model of reference-dependent preferences. *Quarterly Journal of Economics 121*(4), 1133–1165.

Köszegi, S., R. Vetschera, and G. E. Kersten (2004). National cultural differences in the use and perception of internet-based nss: Does high or low context matter? *International Negotiation 9*(1), 79–109.

Kramer, R. M., E. Newton, and P. L. Pommerenke (1993). Self-enhancement biases and negotiator judgment: Effects of self-esteem and mood. *Organizational Behavior and Human Decision Processes 56*(1), 110–133.

Kreps, D. M. (1979). A representation theorem for "preference for flexibility". *Econometrica 47*(3), 565–578.

Kreps, D. M. (1988). *Notes on the Theory of Choice*. Boulder: Westview Press.

Kreps, D. M. (1990). *A Course in Microeconomic Theory*. Harlow: Pearson Education Limited.

Kristensen, H. and T. Gärling (1997a). Adoption of cognitive reference points in negotiations. *Acta Psychologica 97*(3), 277–288.

Kristensen, H. and T. Gärling (1997b). The effects of anchor points and reference points on negotiation processes and outcomes. *Organizational Behavior and Human Decision Processes 71*, 85–94.

Kristensen, H. and T. Gärling (2000). Anchor points, reference points, and couteroffers in negotiations. *Group Decision and Negotiation 9*, 453–505.

Ku, G. (2000). Auctions and auction fever: Explanations from competitive arousal and framing. *Kellog Journal of Organization Behavior*, 1–31.

Ku, G., D. Malhotra, and J. K. Murnighan (2005). Towards a competitive arousal model of decision-making: A study of auction fever in live and internet auctions. *Organizational Behavior and Human Decision Processes 96*(2), 89–103.

Lai, H., H.-S. Doong, C.-C. Kao, and G. E. Kersten (2006). Understanding behavior and perception of negotiators from their strategies. In *Proceedings of the 39th Annual Hawaii International Conference on System Sciences (HICSS'06)*, Los Alamitos. IEEE Computer Society.

Lechner, U. and B. Schmid (1999). Logic for media - the computational media metaphor. In *Proceedings of the 32nd Annual Hawaii International Conference on Systems Sciences (HICSS'99)*, Los Alamitos. IEEE Computer Society.

Lerner, J. S., D. A. Small, and G. F. Loewenstein (2004). Cheart strings and purse strings: Carryover effects of emotions on economic decisions. *Psychological Science 15*(5), 337–341.

Li, D. (2004). Bargaining with history-dependent preferences. Working Paper.

Lichtenstein, S. and P. Slovic (1971). Reversals of preference between bids and choices in gambling decisions. *Journal of Experimental Psychology 59*(1), 46–55.

Lieberman, B. (1962). Experimental studies of conflict in some two-person and three-person games. In J. H. Criswell, H. Solomon, and P. Suppes (Eds.), *Mathematical Models in Small Group Processes*, pp. 203–220. Stanford: Stanford University Press.

Lim, R. (1997). Overconfidence in negotiation revisited. *International Journal of Conflict Management 8*(1), 52–79.

Lindman, H. R. (1971). Inconsistent preferences among gambles. *Journal of Experimental Psychology 80*(2), 590–597.

List, J. A. (2001). Do explicit warnings eliminate the hypothetical bias in elicitation procedures? evidence from field auctions for sportscards. *American Economic Review 91*(4), 1498–1507.

List, J. A. (2003). Does market experience eliminate market anomalies. *The Quarterly Journal of Economics 118*(1), 41–71.

List, J. A. (2004). Neoclassical theory versus prospect theory: Evidence from the marketplace. *Econometrica 72*(2), 615–625.

List, J. A. and D. Lucking-Reiley (2000). Demand reduction in mulitunit auctions: Evidence from a sportscard field experiment. *American Economic Review 90*(4), 961–972.

Little, R. J. A. (1989). Testing the equality of two independent binomial proportions. *The American Statistician 43*, 283–288.

Loewenstein, G. F. and S. Issacharoff (1994). Source dependence in the valuation of objects. *Journal of Behavioral Decision Making 7*, 157–168.

Loewenstein, G. F., S. Issacharoff, and C. Camerer (1993). Self-serving assessments of fairness and pretrial bargaining. *Journal of Economic Behavior and Organization 22*, 135–159.

Loewenstein, G. F. and T. O'Donoghue (2005). Animal spirits: Affective and deliberative processes in economic behavior. Working Paper.

Logothetis, N. K., J. Pauls, M. Augath, T. Trinath, and A. Oeltermann (2001). Neurophysiological investigation of the basis of the fmri signal. *Nature 12*(412), 150–157.

Loomes, G. and R. Sudgen (1982). Regret theory: An alternative theory of rational choice under uncertainty. *The Economic Journal 92*(368), 805–824.

Lucking-Reiley, D. H. (1999). Using field experiments to test equivalence between auction formats: Magic on the internet. *American Economic Review 89*(5), 1063–1080.

Mann, H. and D. Whitney (1947). On a test of whether one of two random variables is stochastically larger than the other. *Annals of Mathematical Statistics 18*, 50–60.

March, J. G. (1978). Bounded rationality, ambiguity, and the engineering of choice. *Bell Journal of Economics 9*(2), 597–608.

March, J. G. and H. A. Simon (1958). *Organizations*. New York: John Wiley & Sons.

Marshall, A. (1997). *Principles of Economics*. Prometheus Books. Original 1890.

Mas-Colell, A., M. D. Whinston, and J. R. Green (1995). *Microeconomic Theory*. Oxford: Oxford University Press.

McAfee, R. P. and J. McMillan (1987). Auctions and bidding. *Journal of Economic Literature 25*(2), 699–738.

McAfee, R. P. and J. McMillan (1996). Analyzing the airwaves auction. *Journal of Economic Perspectives 10*, 159–175.

McCabe, K., D. Houser, L. Ryan, V. L. Smith, and T. Trouard (2001). A functional imaging study of cooperation in two-person reciprocal exchange. *Proceedings of the National Academy of Sciences 98*(20), 11832–11835.

McCabe, K. A. (2003). Neuroeconomics. In L. Nadel (Ed.), *Encyclopedia of Cognitive Science*, Volume 3, pp. 294–298. London: Nature Publishing Group, Macmillan Publishers Ltd.

McCloskey, D. N. (1994). *Knowledge and Persuasion in Economics.* Cambridge: Cambridge University Press.

McClure, S. M., D. I. Laibson, G. F. Loewenstein, and J. D. Cohen (2004). Seperate neural systems value immediate and delayed monetary rewards. *Science 306*(5695), 503–507.

Messick, D. M. (1967). Interdependent decision strategies in zero-sum games: A computer controlled study. *Behavioral Science 12*, 33–48.

Morrison, G. C. (1997). Willingness to pay and willingness ot accept: Some evidence of an endowment effect. *Applied Economics 29*(4), 411–417.

Musch, J. and K. C. Klauer (2002). Psychological experimenting on the world wide web: Investigating content effects in syllogistic reasoning. In B. Batinic, U.-D. Reips, and M. Bosnjak (Eds.), *Onlines Social Sciences*, Chapter 11, pp. 181–212. Hogrefe & Huber Publishers.

Musch, J. and U.-D. Reips (2000). A brief history of web experimenting. In M. H. Birnbaum (Ed.), *Psychological experiments on the Internet*, pp. 61–85. San Diego: Academic Press.

Myerson, R. B. and M. A. Satterthwaite (1983). Efficient mechanisms for bilateral trading. *Journal of Economic Theory 29*(2), 265–281.

Nash, J. F. (1950). The bargaining problem. *Econometrica 18*(2), 155–162.

Nash, J. F. (1951). Non-cooperative games. *The Annals of Mathematics 54*(2), 286–295.

Nash, J. F. (1953). Two-person cooperative bargaining games. *Econometrica 21*(1), 128–140.

Neale, M. (1984). The effects of negotiation and arbitration cost salience on bargainer behavior: the role of the arbitrator and constituency on negotiator judgment. *Organizational Behavior and Human Decision Processes 34*, 97–111.

Neale, M. A. and M. H. Bazerman (1991). *Cognition and Rationality in Negotiation.* New York: Free Press.

Neumann, D. (2004). *Market Engineering - A Structured Design Process for Electronic Markets.* Ph. D. thesis, School of Economcis and Business Engineering, University of Karlsruhe, Germany.

Northcraft, G. and M. Neale (1986). Opportunity costs and the framing of resource allocation decisions. *Organizational Behavior and Human Decision Processes 37*, 348–356.

Northcraft, G. and M. Neale (1987). Expert, amateurs, and real estate: An anchoring-and-adjustment perpective on property pricing decisions. *Organizational Behavior and Human Decision Processes 39*, 228–241.

Novemsky, N. and D. Kahneman (2005). The boundaries of loss aversion. *Journal of Marketing Research 42*(May), 119–128.

Nowlis, S. M. and I. Simonson (1997). Attribute-task compatibility as a determinnat of consumer preference reversals. *Journal of Marketing Research 34*(2), 205–218.

O'Donoghue, T. and M. Rabin (1999). Doing it now or later. *American Economic Review 89*(1), 103–124.

Olekalns, M. (1997). Situational cues as moderators of the frame-outcome relationship. *British Journal of Social Psychology 36*(2), 191–209.

Parsons, L. M. and D. Osherson (2001). New evidence for distinct right and left brain systems for deductive versus probabilistic reasoning. *Cerebral Cortex 11*(10), 954–965.

Patefield, W. M. (1981). An efficient method of generating r x c tables with given row and column totals (algorithm as159). *Applied Statistics 30*(1), 91–97.

Payne, J. W., J. R. Bettman, and E. J. Johnson (1993). *The Adaptive Decision Maker*. Cambridge: Cambridge University Press.

Pearson, K. (1896). Mathematical contributions to the theory of evolution—III. regression, heredity and panmixia. *Philosophical Transactions of the Royal Society of London 187*, 253–318.

Pearson, K. (1900). On the criterion that a given system of deviations from the probable in the case of a correlated system of variables is such that it can reasonably be supposed to have arisen in a random sampling. *Philosophical Magazine 5*, 157–175.

Pinkley, R., T. Griffith, and G. Northcraft (1995). Fixed pie a la mode: information availability, information processing, and the negotiation of suboptimal agreements. *Organizational Behavior and Human Decision Processes 62*, 101–112.

Pitman, E. J. G. (1937). Significance tests which may be applied to samples from any population (parts I and II). *Royal Statistical Society Supplement 4*, 119–130 and 225–232.

Pitman, E. J. G. (1938). Significance tests which may be applied to samples from any population. Part III. the analysis of variance test. *Biometrika 29*, 322–335.

Plott, C. R. (1996). Rational individual behavior in markets and social choice processes: The discovered preference hypothesis. In K. J. Arrow, E. Colombatto, M. Perlamn, and C. Schmidt (Eds.), *The Rational Foundations of Economic Behavior*, pp. 225–250. London, New York: Macmillan Press Ltd. and St. Martin's Press Inc.

Plott, C. R. and K. Zeiler (2005). The willingness to pay-willingness to accept gap, the "endowment effect," subject misconceptions, and experimental procedures for eliciting valuations. *American Economic Review 95*(3), 530–545.

Prelec, D. (1998). The probability weighting function. *Econometrica 66*(3), 497–527.

Pruitt, D. G. and P. J. Carnevale (1993). *Negotiation in Social Conflict*. Pacific Grove: Brooks/Cole.

Quandt, R. E. (1956). A probabilistic theory of consumer behavior. *The Quarterly Journal of Economics 70*(4), 507–536.

Quiggin, J. (1982). A theory of anticipated utility. *Journal of Economic Behavior and Organization 3*(4), 323–343.

R Development Core Team (2005). *R: A language and environment for statistical computing*. Vienna, Austria: R Foundation for Statistical Computing. ISBN 3-900051-07-0.

Raiffa, H. (1982). *The Art and Science of Negotiation*. Cambridge: Harvard University Press.

Raiffa, H. (2003). *Negotiation Analysis: The Science and Art of Collaborative Decision Making*. Cambridge: Harvard University Press. With contributing authors D. Metcalfe and J. Richardson.

Read, D. (2005). Monetary incentives, what are they good for? *Journal of Economic Methodology 12*(2), 265–276.

Reiley, D. H. (2006). Field experiments on the effects of reserve prices in auctions: More magic on the internet. *RAND Journal of Economics* (1), 195–211.

Reips, U.-D. (2000). The web experiment method: Advantages, disadvantages, and solutions. In M. H. Birnbaum (Ed.), *Psychological experiments on the Internet*, pp. 89–114. San Diego: Academic Press.

Reips, U.-D. (2002a). Standards for internet-based experimenting. *Experimental Psychology 49*(4), 243–256.

Reips, U.-D. (2002b). Theory and techniques of conducting web experiments. In B. Batinic, U.-D. Reips, and M. Bosnjak (Eds.), *Onlines Social Sciences*, Chapter 13, pp. 229–250. Seattle: Hogrefe & Huber Publishers.

Rilling, J. K., D. A. Gutman, T. R. Zeh, G. Pagnoni, G. S. Berns, and C. D. Kilts (2002). A neural basis for social cooperation. *Neuron 35*, 395–405.

Ritov, I. (1996). Anchoring in simulated competitive market negotiation. *Organizational Behavior and Human Decision Processes 67*, 16–25.

Ross, L. and C. Stillinger (1991). Barriers to conflict resolution. *Nego-tiation Journal 7*, 398–404.

Roth, A. E. (Ed.) (1985). *Game-Theoretic Models of Bargaining*. Cambrdige: Cambridge University Press.

Roth, A. E. (1996). Comments on tversky's 'rational theory and constructive choice'. In K. Arrow, E. Colombatto, M. Perlman, and C. Schmidt (Eds.), *Foundations of Economic Behavior*, pp. 198–202. London: Macmillan.

Roth, A. E. (2002). The economist as an engineer: Game theory, experimentation, and computation as tools for design economics. *Econometrica 70*(4), 1341–1378.

Roth, A. E. and E. Pearson (1999). The redesign of the matching market for american physicians: Some engineering aspects of economic design. *American Economic Review 89*, 748–780.

Roth, A. E. and F. Schoumaker (1983). Expectations and reputations in bargaining: An experimental study. *American Economic Review 73*(3), 362–372.

Rubinstein, A. (1982). Perfect equilibrium in a bargaining model. *Econometrica 50*(1), 97–110.

Rubinstein, A. (1985). A bargaining model with incomplete information about time preferences. *Econometrica 53*(5), 1151–1172.

Rubinstein, A. (1988). Similarity and decision-making under risk. *Journal of Economic Theory 46*(1), 145–153.

Rubinstein, A. (2001). A theorist's view of experiments. *European Economic Review 45*(3), 615–628.

Russo, J. E. (1977). The value of unit price information. *Journal of Marketing Research 14*(2), 193–201.

Rutström, E. E. (1998). Home-grown values and the design of incentive compatible auctions. *International Journal of Game Theory 27*(3), 427–441.

Saaty, T. L. (1980). *Multicriteria Decision Making: The Analytic Hierarchy Process*. New York: McGraw-Hill.

Samuelson, W. and R. Zeckhauser (1981). Status quo bias in decision making. *Journal of Risk and Uncertainty 1*(1), 7–59.

Sanfey, A. G., J. K. Rilling, J. A. Aronson, L. E. Nystrom, and J. D. Cohen (2003). The neural basis of economic decision-making in the ultimatum game. *Science 300*, 1755–1758.

Schmeidler, D. (1989). Subjective probability and expected utility without additivity. *Econometrica 57*(3), 571–587.

Schmid, B. F. (1999). Elektronische märkte - merkmale, organisation und potentiale. In A. Hermanns and M. Sauter (Eds.), *Management-*

Handbuch Electronic Commerce, pp. 31–48. München: Verlag Franz Vahlen GmbH.

Schoop, M., F. Köhne, and D. Staskiewicz (2004). An integrated decision and communication perspective on electronic negotiation support systems: Challenges and solutions. *Journal of Decision Systems 13*(4), 375–398.

Seifert, S. (2005). *Posted Price Offers in Internet Auction Markets*. Ph. D. thesis, School of Economcis and Business Engineering, University of Karlsruhe, Germany.

Selten, R. (1967). Die Strategiemethode zur Erforschung des eingeschränkt rationalen Verhaltens im Rahmen eines Oligopolexperiments. In H. Sauermann (Ed.), *Beiträge zur experimentellen Wirtschaftsforschung, 1. Band*, pp. 136–168. Tübingen: J. C. B. Mohr (Paul Siebeck).

Selten, R. (2001). What is bounded rationality? In G. Gigerenzer and R. Selten (Eds.), *Bounded Rationality: The Adaptive Toolbox*, pp. 13–36. Cambridge: MIT Press.

Shachat, J. and J. T. Swarthout (2002). Learning about learning in games through experimental control of strategic interdependence. Working Paper.

Shachat, J. and J. T. Swarthout (2004). Do we detect and exploit mixed strategy play by opponents? *Mathematical Methods of Operations Research 59*(3), 359–373.

Shalev, J. (2002). Loss aversion and bargaining. *Theory and Decision 52*, 201–232.

Sheskin, D. J. (2004). *Handbook of parametric and nonparametric statistical procedures* (3. ed.). Boca Raton: Chapman & Hall/CRC.

Shogren, J. F., S. Y. Shin, D. J. Hayes, and J. B. Kliebenstein (1994). Resolving differences in willingness to pay and willingness to accept. *American Economic Review 84*(1), 255–270.

Simon, H. A. (1955). A behavioral model of rational choice. *Quarterly Journal of Economics 69*(1), 99–118.

Simon, H. A. (1957). *Models of Man*. New York: John Wiley & Sons.

Simon, H. A. (1973). The structure of ill-structured problems. *Artificial Intelligence 4*, 181–201.

Slovic, P. (1975). Choice between equally valued alternatives. *Journal of Experimental Psychology: Human Perception and Performance 1*(3), 280–287.

Slovic, P. (1995). The construction of preferences. *American Psychologist 50*(5), 364–371.

Slovic, P., D. Griffin, and A. Tversky (2002). Compatibility effects in judgment and choice. In T. Gilovich, D. Griffin, and D. Kahneman (Eds.), *Heuristics and Biases: The Psychology of Intuitive Judgment*, pp. 217–229. Cambridge: Cambridge University Press.

Smith, K., J. Dickhaut, K. McCabe, and J. V. Pardo (2002). Neuronal substrates for choice und ambiguity, risk, gains, and losses. *Management Science 48*(6), 711–718.

Smith, M. and B. Leigh (1997). Virtual subjects: Using the internet as an alternative source of subjects and research environment. *Behavior Research Methods, Instruments, and Computers 29*, 496–505.

Smith, V. L. (1976a). Bidding and auctioning institutions: Experimental results. In Y. Amihud (Ed.), *Bidding and Auctioning for Procurement and Allocation*, pp. 43–64. New York: New York University Press.

Smith, V. L. (1976b). Experimental economics: Induced value theory. *American Economic Review 66*(2), 274–279.

Smith, V. L. (1982a). Microeconomic systems as an experimental science. *American Economic Review 72*(5), 923–955.

Smith, V. L. (1982b). Reflections on some experimental market mechanisms for classical environments. In L. McAllister (Ed.), *Choice Models for Buyer Behavior*, pp. 13–47. Greenwich: JAI Press.

Smith, V. L. (2003). Constructivist and ecological rationality in economics. In T. Frängsmyr (Ed.), *Les prix Nobel 2002: Nobel prizes, presentations, biographies, and lectures.*, pp. 502–561. Stockholm: Almqvist & Wiksell International.

Smith, V. L. and J. M. Walker (2000). Monetary rewards and decision cost in experimental economics. In V. L. Smith (Ed.), *Bargaining and Market Behavior*, pp. 41–60. Cambridge: Cambridge University Press.

Spearman, C. (1904). The proof and measurement of association between two things. *American Journal of Psychology 15*, 72–101.

Starmer, C. (2000). Developments in non-expected utility theory: The hunt for a descriptive theory of choice under risk. *Journal of Economic Literature 38*(2), 332–382.

Stigler, G. J. (1950). The development of utility theory. II. *The Journal of Political Economy 58*(5), 373–396.

Stigler, G. J. and G. S. Becker (1977). De gustibus non est disputandum. *American Economic Review 67*(2), 76–90.

Strahilevitz, M. A. and G. F. Loewenstein (1998). The effect of ownership history on the valuation of objects. *Journal of Consumer Research 25*, 276–289.

Ströbel, M. (2003). *Engineering Electronic Negotiations*. Series in Computer Science. Dordrecht: Kluwer Academic Publishers.

Ströbel, M. and C. Weinhardt (2003). The montreal taxonomy for electronic negotiations. *Group Decision and Negotiation Journal 12*(2), 143–164.

Thaler, R. H. (1980). Towards a positive theory of consumer choice. *Journal of Economic Behavior and Organization 1*(March), 39–60.

Thiessen, E. M. and A. Soberg (2003). Smartsettle described with the montreal taxonomy. *Group Decision and Negotiation 12*(2), 165–170.

Thompson, L. L. (1990). Negotiation behavior and outcomes: empirical evidence and theoretical issues. *Psychological Bulletin 108*, 515–532.

Thompson, L. L. (1995). The impact of minimum goals and aspirations on judgments of success in negotiations. *Group Decision and Negotiation 4*, 513–524.

Thompson, L. L. and T. DeHarpport (1994). Social judgment, feedback, and interpersonal learning in negotiation. *Organizational Behavior and Human Decision Processes 58*, 327–345.

Thompson, L. L. and R. Hastie (1990). Social perception in negotiation. *Organizational Behavior and Human Decision Processes 47*, 98–123.

Thompson, L. L. and D. Hrebec (1996). Lose-lose agreements in interdependent decision-making. *Psychological Bulletin 120*, 396–409.

Thomson, W. (1994). Cooperative models of bargaining. In R. Aumann and S. Hart (Eds.), *Handbook of Game Theory with Economic Applications*, Volume 2, Chapter 35, pp. 1237–1284. Amsterdam: North-Holland.

Turel, O. (2006). Actor-partner effects in e-negotiation: Extending the assessment model of internet systems. In S. Seifert and C. Weinhardt (Eds.), *Group Decision and Negotiation (GDN) 2006*, pp. 238–240. Karlsruhe: Universitätsverlage Karlsruhe.

Tverksy, A. and D. Kahneman (1974). Judgement under uncertainty: Heuristics and biases. *Science 185*, 1124–1131.

Tversky, A. (1972). Elimination by aspects: A theory of choice. *Psychological Review 79*(4), 281–299.

Tversky, A. and D. Kahneman (1981). The framing of decisions and the psychology of choice. *Science 211*(4481), 453–458.

Tversky, A. and D. Kahneman (1986). Rational choice and the framing of decisions. *The Journal of Business 59*(4), 251–278.

Tversky, A. and D. Kahneman (1991). Loss aversion in riskless choice: A reference-dependent model. *Quarterly Journal of Economics 106*(4), 1039–1061.

Tversky, A. and D. Kahneman (1992). Advances in prospect theory: Cumulative representation of uncertainty. *Journal of Risk and Uncertainty 5*(4), 297–323.

Tversky, A., S. Sattath, and P. Slovic (1988). Contingent weighting in judegemnt and choice. *Psychological Review 95*(3), 371–384.

Tyebjee, T. T. (1979). Response time, conflict and involvement in brand choice. *Journal of Consumer Research 6*(3), 295–304.

Valley, K., J. Moag, and M. Bazerman (1998). A matter of trust: effects of communication on the efficiency and distribution of outcomes. *Journal of Economic Behavior and Organization 34*, 211–238.

Varian, H. R. (1992). *Microeconomic Analysis* (3. ed.). New York: W. W. Norton & Company.

Varian, H. R. (2002). When economics shifts from science to engineering. *New York Times* (August 29, 2002).

Vetschera, R. (2004a). Estimating negotiator performance without preference information. Working Paper.

Vetschera, R. (2004b). Preference structures and behavioral consistency in negotiations. Working Paper.

Viscusi, W. K., W. A. Magat, and J. Huber (1987). An investigation of the rationality of consumer valuations of multiple health risks. *RAND Journal of Economics 18*(4), 465–479.

von Neuman, J. and O. Morgenstern (1944). *The Theory of Games and Economic Behavior* (1. ed.). Princeton: Princeton University Press.

von Neuman, J. and O. Morgenstern (1947). *The Theory of Games and Economic Behavior* (2. ed.). Princeton: Princeton University Press.

Walker, J., V. Smith, and J. Cox (1987). Bidding behavior in first price sealed bid auctions: Use of computerized nash competitors. *Economics Letters 23*(3), 239–244.

Weinhardt, C. and H. Gimpel (2006). Market engineering: An interdisciplinary research challenge. In N. Jennings, G. Kersten, A. Ockenfels, and C. Weinhardt (Eds.), *Negotiation and Market Engineering*, Number 06461 in Dagstuhl Seminar Proceedings. Internationales Begegnungs- und Forschungszentrum (IBFI), Schloss Dagstuhl, Germany.

Weinhardt, C., C. Holtmann, and D. Neumann (2003). Market engineering. *Wirtschaftsinformatik 45*(6), 635–640.

Weinhardt, C., D. Neumann, and C. Holtmann (2006). Computer-aided market engineering. *Communications of the ACM 49*(7), 79.

Wessells, M. G. (1982). *Cognitive Psychology*. New York: Harper and Row Publishers Inc.

Weymark, J. A. (1981). Generalized gini inequality indices. *Mathematical Social Sciences 1*(4), 409–430.

White, S. B., K. L. Valley, M. H. Bazerman, M. A. Neale, and S. R. Peck (1994). Alternative models of price behavior in dyadic negotiations: Market prices, reservation prices, and negotiator aspirations. *Organizational Behavior and Human Decision Processes 57*, 430–447.

Whyte, G. and J. K. Sebenius (1997). The effect of multiple anchors on anchoring in individual and group judgment. *Organizational Behavior and Human Decision Processes 69*, 75–85.

Wilcoxon, F. (1949). *Introduction to robust ersitmation and hypothesis testing.* San Diego: Academic Press.

Wilde, L. L. (1980). On the use of laboratory experiment in economics. In J. C. Pitt (Ed.), *Philospohy in Economics.* Dordrecht: Reidel.

Wilson, R. (2002). Architecture of power markets. *Econometrics 70*, 1299–1340.

Winter, E. and S. Zamir (2005). An experiment with ultimatum bargaining in a changing environment. *Japanese Economic Review 56*(3), 363–385.

Yaari, M. E. (1987). The dual theory of choice under risk. *Econometrica 55*(1), 95–115.

Yates, F. (1934). Contingency tables involving small numbers and the chi-square test. *Journal of the Royal Statistical Society 1*, 217–235.

Yates, F. (1984). Test of significance for 2×2 contingency tables (with discussion). *Journal of the Royal Statistical Society. Series A 147*, 426–463.

List of Abbreviations and Symbols

Abbreviations

AHP	Analytical Hierarchy Process
BATNA	Best alternative to a negotiated agreement
CD	Compact Disk
cf.	confer
Ch.	Chapter
EEG	Electro Encephalogram
e.g.	for example ('exempli gratia')
esp.	especially
Fig.	Figure
FCC	Federal Communications Commission
fMRI	functional Magnetic Resonance Imaging
HTML	Hypertext Markup Language
i.e.	that is ('id est')
IP	Internet Protocol
IT	Information Technology
NIRS	Near Infrared Spectography
p. / pp.	page / pages
PET	Positron Emission Tomography
Sec.	Section
sim.	simulated
SMART	Simple Multi-attribute Rating Technique
SMARTER	SMART Exploiting Ranks
US	United States of America
w.l.o.g.	without loss of generality
WTA	Willingness to accept
WTP	Willingness to pay

Symbols

$\succ$	strict preference
$\succeq$	weak preference
$\sim$	indifference
$\in$	set membership
$\subseteq$	set inclusion (includes set equality)
$\exists$	existential quantifier
$\nexists$	non-existential quantifier
$\forall$	allquantor (for all)
$\longmapsto$	causal relationship
$\Rightarrow$	implication
$\Leftrightarrow$	equivalence
$\wedge$	logical and
$\vee$	logical or
$\neg$	negation
$\mathbb{R}$	set of real numbers
$\mathbb{R}_+$	set of positive real numbers including zero
$\langle \cdots \rangle$	specification of a vector
$\{ \cdots \}$	specification of a set
$\#$	number of

Lecture Notes in Economics and Mathematical Systems

For information about Vols. 1–505
please contact your bookseller or Springer-Verlag

Vol. 549: G. N. Krieg, Kanban-Controlled Manufacturing Systems. IX, 236 pages. 2005.

Vol. 550: T. Lux, S. Reitz, E. Samanidou, Nonlinear Dynamics and Heterogeneous Interacting Agents. XIII, 327 pages. 2005.

Vol. 551: J. Leskow, M. Puchet Anyul, L. F. Punzo, New Tools of Economic Dynamics. XIX, 392 pages. 2005.

Vol. 552: C. Suerie, Time Continuity in Discrete Time Models. XVIII, 229 pages. 2005.

Vol. 553: B. Mönch, Strategic Trading in Illiquid Markets. XIII, 116 pages. 2005.

Vol. 554: R. Foellmi, Consumption Structure and Macroeconomics. IX, 152 pages. 2005.

Vol. 555: J. Wenzelburger, Learning in Economic Systems with Expectations Feedback (planned) 2005.

Vol. 556: R. Branzei, D. Dimitrov, S. Tijs, Models in Cooperative Game Theory. VIII, 135 pages. 2005.

Vol. 557: S. Barbaro, Equity and Efficiency Considerations of Public Higer Education. XII, 128 pages. 2005.

Vol. 558: M. Faliva, M. G. Zoia, Topics in Dynamic Model Analysis. X, 144 pages. 2005.

Vol. 559: M. Schulmerich, Real Options Valuation. XVI, 357 pages. 2005.

Vol. 560: A. von Schemde, Index and Stability in Bimatrix Games. X, 151 pages. 2005.

Vol. 561: H. Bobzin, Principles of Network Economics. XX, 390 pages. 2006.

Vol. 562: T. Langenberg, Standardization and Expectations. IX, 132 pages. 2006.

Vol. 563: A. Seeger (Ed.), Recent Advances in Optimization. XI, 455 pages. 2006.

Vol. 564: P. Mathieu, B. Beaufils, O. Brandouy (Eds.), Artificial Economics. XIII, 237 pages. 2005.

Vol. 565: W. Lemke, Term Structure Modeling and Estimation in a State Space Framework. IX, 224 pages. 2006.

Vol. 566: M. Genser, A Structural Framework for the Pricing of Corporate Securities. XIX, 176 pages. 2006.

Vol. 567: A. Namatame, T. Kaizouji, Y. Aruga (Eds.), The Complex Networks of Economic Interactions. XI, 343 pages. 2006.

Vol. 568: M. Caliendo, Microeconometric Evaluation of Labour Market Policies. XVII, 258 pages. 2006.

Vol. 569: L. Neubecker, Strategic Competition in Oligopolies with Fluctuating Demand. IX, 233 pages. 2006.

Vol. 570: J. Woo, The Political Economy of Fiscal Policy. X, 169 pages. 2006.

Vol. 571: T. Herwig, Market-Conform Valuation of Options. VIII, 104 pages. 2006.

Vol. 572: M. F. Jäkel, Pensionomics. XII, 316 pages. 2006

Vol. 573: J. Emami Namini, International Trade and Multinational Activity. X, 159 pages, 2006.

Vol. 574: R. Kleber, Dynamic Inventory Management in Reverse Logistics. XII, 181 pages, 2006.

Vol. 575: R. Hellermann, Capacity Options for Revenue Management. XV, 199 pages, 2006.

Vol. 576: J. Zajac, Economics Dynamics, Information and Equilibnum. X, 284 pages, 2006.

Vol. 577: K. Rudolph, Bargaining Power Effects in Financial Contracting. XVIII, 330 pages, 2006.

Vol. 578: J. Kühn, Optimal Risk-Return Trade-Offs of Commercial Banks. IX, 149 pages, 2006.

Vol. 579: D. Sondermann, Introduction to Stochastic Calculus for Finance. X, 136 pages, 2006.

Vol. 580: S. Seifert, Posted Price Offers in Internet Auction Markets. IX, 186 pages, 2006.

Vol. 581: K. Marti; Y. Ermoliev; M. Makowsk; G. Pflug (Eds.), Coping with Uncertainty. XIII, 330 pages, 2006.

Vol. 582: J. Andritzky, Sovereign Default Risks Valuation: Implications of Debt Crises and Bond Restructurings. VIII, 251 pages, 2006.

Vol. 583: I.V. Konnov, D.T. Luc, A.M. Rubinov[†] (Eds.), Generalized Convexity and Related Topics. IX, 469 pages, 2006.

Vol. 584: C. Bruun, Adances in Artificial Economics: The Economy as a Complex Dynamic System. XVI, 296 pages, 2006.

Vol. 585: R. Pope, J. Leitner, U. Leopold-Wildburger, The Knowledge Ahead Approach to Risk. XVI, 218 pages, 2007 (planned).

Vol. 586: B.Lebreton, Strategic Closed-Loop Supply Chain Management. X, 150 pages, 2007 (planned).

Vol. 587: P. N. Baecker, Real Options and Intellectual Property: Capital Budgeting Under Imperfect Patent Protection. X, 276 pages , 2007.

Vol. 588: D. Grundel, R. Murphey, P. Panos , O. Prokopyev (Eds.), Cooperative Systems: Control and Optimization. IX, 401 pages , 2007.

Vol. 589: M. Schwind, Dynamic Pricing and Automated Resource Allocation for Information Services: Reinforcement Learning and Combinatorial Auctions. XII, 293 pages , 2007.

Vol. 590: S. H. Oda, Developments on Experimental Economics: New Approaches to Solving Real-World Problems. XVI, 262 pages, 2007.

Vol. 591: M. Lehmann-Waffenschmidt, Economic Evolution and Equilibrium: Bridging the Gap. VIII, 272 pages, 2007.

Vol. 592: A. C.-L. Chian, Complex Systems Approach to Economic Dynamics. X, 95 pages, 2007.

Vol. 593: J. Rubart, The Employment Effects of Technological Change: Heterogenous Labor, Wage Inequality and Unemployment. XII, 209 pages, 2007

Vol. 594: R. Hübner, Strategic Supply Chain Management in Process Industries: An Application to Specialty Chemicals Production Network Design. XII, 243 pages, 2007

Vol. 595: H. Gimpel, Preferences in Negotiations: The Attachment Effect. XIV, 268 pages, 2007

Vol. 596: M. Müller-Bungart, Revenue Management with Flexible Products: Models and Methods for the Broadcasting Industry.XXI, 297 pages, 2007

Printing: Krips bv, Meppel
Binding: Stürtz, Würzburg